INTERNATIONAL CENTRE FOR MECHANICAL SCIENCES

COURSES AND LECTURES - No. 268

NONLOCAL THEORY OF MATERIAL MEDIA

EDITED BY

D. ROGULA

POLISH ACADEMY OF SCIENCES

SPRINGER-VERLAG WIEN GMBH

This volume contains 27 figures.

Originally published by Springer-Verlag Wien New York in 1982

ISBN 978-3-211-81632-5 ISBN 978-3-7091-2890-9 (eBook)
DOI 10.1007/978-3-7091-2890-9

PREFACE

Although the idea of interaction distance appears as old as human knowledge, the Nonlocal Theory of Material Media is a relatively recent discipline. The fact that this theory has not been developed until recently can perhaps be explained by the great success both of the classical continuum theory with contact interactions and, on the other hand, of the statistical molecular mechanics. However, the growing need for understanding the phenomena intermediate between these two extremities promoted the development of the theory of material media, based on nonlocal concepts. This theory bridges the continuum mechanics and the molecular physics by a general representation of interaction force and kinematical properties of condensed matter. The classical models of solids, i.e. the classical continuum with its contact forces and a discrete lattice with interaction of molecules distant from each other, may be regarded as the extreme instances of nonlocal material media.

In spite of the fact that at present the theory is far from being complete, a number of ideas have developed which can be of broad interest, even for the non-specialist. The present volume contains essentially the lectures on Nonlocal Theory of Material Media given at the Centre International des Sciences Mécaniques. The aim of the volume is to sketch the physical and mathematical foundations of the nonlocal theory of material media, its general results, applications, connexions with related domains of mechanics, and many questions open for future research. Special attention is paid to the problems of structural defects and boundaries of solids.

It is a pleasure to acknowledge the Centre International des Sciences Mécaniques for its helpfulness. Our gratitude is also due to the authors who contributed to this volume.

Dominik Rogula

Warsaw, June 1982

CONTENTS

Page

Page

DEFECTS IN CRYSTALLINE MEDIA

A.G. Crocker
Department of Physics
University of Surrey
Guildford GU2 5XH
England

Perfect and imperfect crystals

Introduction. Various theoretical methods of investigating the properties of defects in crystalline media will be explored in this chapter. In particular emphasis will be placed on methods based on discrete models using inter-atomic potentials and computer simulation. However these methods rely heavily on earlier techniques based on geometrical and continuum models, which are also described. All models of defects must of course be developed from an understanding of the properties of perfect crystals. Therefore in the present section some of the relevant basic definitions are summarised. The types of defect which can arise in crystals are then discussed in general terms and the main experimental techniques which have been used to provide information on defects are outlined. Subsequent sections deal with geometrical, continuum and discrete theories of defects and finally various relationships between the theories are discussed.

Lattice geometry. Crystals consist of atoms, ions or molecules at the points of lattices, which are infinite arrays of points each point having identical neighbours.[1-3] A particular lattice is defined by a unit cell which is a parallelepiped formed by three pairs of parallel planes, each passing through three non-collinear points. Three concurrent edges of such a cell define a possible direct lattice basis, $\underset{\sim}{a}_i$ (i = 1,2,3). A primitive unit cell is associated with just one lattice point. An infinite number of these unit cells is possible, the one chosen in practice being that with the shortest edges and/or with angles close to $\frac{\pi}{2}$. In practice it is often convenient to choose non-primitive unit cells with two or more points in order to take advantage of orthogonal bases which then become possible. In particular body-centred, face-centred and base-centred cells are used.

A lattice direction is any direction joining two lattice points and is defined by $u^i\underset{\sim}{a}_i$, where u^i are integers.[4] Conventionally the components or indices u^i are written $[u^i]$. The length of the vector $u^i\underset{\sim}{a}_i$ is $(u^iu^jg_{ij})^{\frac{1}{2}}$ where $g_{ij} = \underset{\sim}{a}_i.\underset{\sim}{a}_j$ is the metric tensor of the direct lattice. The angle between two directions $u^i\underset{\sim}{a}_i$ and $v^j\underset{\sim}{a}_j$ is θ where

$$\cos\theta = (u^iv^jg_{ij})\ (u^iu^jg_{ij})^{-\frac{1}{2}}\ (v^iv^jg_{ij})^{-\frac{1}{2}}\quad .$$

These directions are orthogonal when $u^iv^jg_{ij} = 0$.

A lattice plane is any plane passing through three non-collinear lattice points and may be defined by the vector $h^i\underset{\sim}{a}_i$ normal to itself.

However in general h^i will not then be integers and it is therefore convenient to choose a new basis $\underset{\sim}{a}^i$ such that the normal is $h_i\underset{\sim}{a}^i$ where h_i are integers. An infinite number of choices for $\underset{\sim}{a}^i$, the reciprocal lattice basis, is possible. If h_i have no common factor they are known as Miller indices and are conventionally written (h_i). The length of the vector $h_i\underset{\sim}{a}^i$, which is the reciprocal of the interplanar spacing, is $(h_i h_j g^{ij})^{\frac{1}{2}}$ where $g^{ij} = \underset{\sim}{a}^i.\underset{\sim}{a}^j$ is the metric tensor of the reciprocal lattice. The angle between the two planes $h_i\underset{\sim}{a}^i$ and $k_j\underset{\sim}{a}^j$ is then ϕ where

$$\cos\phi = (h_i k_j g^{ij})\ (h_i h_j g^{ij})^{-\frac{1}{2}}\ (k_i k_j g^{ij})^{-\frac{1}{2}} \quad .$$

These planes are orthogonal when $h_i k_j g^{ij} = 0$.

The angle between a direction $u^i\underset{\sim}{a}_i$ and a plane normal $h_j\underset{\sim}{a}^j$ is ψ, where

$$\cos\psi = [u^i h_j(\underset{\sim}{a}_i.\underset{\sim}{a}^j)](u^i u^j g_{ij})^{-\frac{1}{2}}(h_i h_j g^{ij})^{-\frac{1}{2}} \quad .$$

If $\underset{\sim}{a}_i$ and $\underset{\sim}{a}^j$ are corresponding reciprocal lattices $\underset{\sim}{a}_i.\underset{\sim}{a}^j = \delta_i^j$ so that

$$\cos\psi = (u^i h_i)\ (u^i u^j g_{ij})^{-\frac{1}{2}}(h_i h_j g^{ij})^{-\frac{1}{2}} \quad .$$

The direction $u^i\underset{\sim}{a}_i$ lies in the plane $h_i\underset{\sim}{a}^i$ if $u^i h_i = 0$. The plane $h_i\underset{\sim}{a}^i$ contains the two directions $u^j\underset{\sim}{a}_j$ and $v^k\underset{\sim}{a}_k$ if $h_i = \varepsilon_{ijk}u^j v^k$. Similarly the direction $u^i\underset{\sim}{a}_i$ is contained in the two planes $h_i\underset{\sim}{a}^i$ and $k_j\underset{\sim}{a}^j$ if $u^i = \varepsilon^{ijk}h_j k_k$. The direction $u^i\underset{\sim}{a}_i$ is perpendicular to the direction $v^j\underset{\sim}{a}_j$ and lies in the plane $h_k\underset{\sim}{a}^k$ if $u^i = \varepsilon^{ijk}g_{j\ell}v^\ell h_k$. Similarly the plane $h_i\underset{\sim}{a}^i$ contains the direction $u^j\underset{\sim}{a}_j$ and is perpendicular to the plane $h_k\underset{\sim}{a}^k$ if $h_i = \varepsilon_{ijk}g^{j\ell}k_\ell u^k$.

Crystal symmetry. Crystals may have the following six types of symmetry element.[1-3]

1) Centre of symmetry or inversion ($x^i \rightarrow - x^i$).

2) Mirror plane.

3) Glide plane or mirror image plus translation parallel to the plane.

4) n-fold rotation axis or rotation of $2\pi/n$.

5) n-fold screw axis or rotation of $2\pi/n$ plus translation parallel to the axis.

6) n-fold inversion axis or rotation of $2\pi/n$ plus inversion.

The integer n may be 1,2,3,4 or 6. Note that these symmetry elements are not all independent. For example the 2-fold inversion axis is equivalent to a mirror plane.

There are 230 different combinations of the above symmetry elements and these are known as Space Groups. However in order to specify the macroscopic properties of crystals only their orientations and not the relative positions of the elements are needed. The remaining elements and their conventional symbols are then:

1)	Centre of symmetry	$\bar{1}$
2)	Mirror plane	m
3)	1,2,3,4, 6-fold rotation axes	1,2,3,4,6
4)	1,2,3,4, 6-fold inversion axes	$\bar{1},\bar{2},\bar{3},\bar{4},\bar{6}$

There are 32 combinations of the elements in this restricted list and these are known as, Point Groups or Crystal Classes. These are contained as follows in the seven crystal systems:

Triclinic	$1, \bar{1}$
Monoclinic	$2, m, 2/m$
Orthorhombic	$22, mm2, mmm$
Tetragonal	$4, \bar{4}, 4/m, 4mm, \bar{4}2m, 4/mmm$
Cubic	$23, m3, 432, \bar{4}32, m3m$
Rhombohedral	$3, \bar{3}, 32, 3m, \bar{3}m$
Hexagonal	$6, \bar{6}, 6/m, 622, 6mm, \bar{6}m2, 6/mmm$

The minimum symmetries for the crystal systems and the resulting restrictions on the direct lattice basis are thus:

Triclinic	No symmetry except 1-fold axis; $\underset{\sim}{a}_i$.
Monoclinic	One 2-fold axis; $\underset{\sim}{a}_1 \cdot \underset{\sim}{a}_3 = \underset{\sim}{a}_2 \cdot \underset{\sim}{a}_3 = 0$.
Orthorhombic	Three orthogonal 2-fold axes; $\underset{\sim}{a}_i \cdot \underset{\sim}{a}_j = 0 \ (i \neq j)$.
Tetragonal	One 4-fold axis; $\underset{\sim}{a}_i \cdot \underset{\sim}{a}_j = 0 \ (i \neq j)$, $a_1 = a_2$.
Cubic	Four 3-fold axes along cube diagonals; $\underset{\sim}{a}_i \cdot \underset{\sim}{a}_j = a^2 \delta_{ij}$.
Rhombohedral	One 3-fold axis; $\lvert \underset{\sim}{a}_i \rvert = a$, $\underset{\sim}{a}_i \cdot \underset{\sim}{a}_j = a^2 \cos \alpha \ (i \neq j)$.
Hexagonal	One 6-fold axis; $\underset{\sim}{a}_1 \cdot \underset{\sim}{a}_2 = -a^2/2$, $\underset{\sim}{a}_1 \cdot \underset{\sim}{a}_3 = \underset{\sim}{a}_2 \cdot \underset{\sim}{a}_3 = 0$ $\lvert \underset{\sim}{a}_1 \rvert = \lvert \underset{\sim}{a}_2 \rvert = a$, $\lvert \underset{\sim}{a}_3 \rvert = c$.

By introducing centred unit cells, higher symmetries can be used. Thus the orthorhombic crystal system can be body-, face- and base-centred, cubic can be body- and face-centred, and monoclinic and tetragonal can be body-centred. Including the seven primitive cells this gives rise to the 14 Bravais Space Lattices.

As a result of crystal symmetry families of up to 48 crystallographically equivalent directions and planes may arise. Individual planes are conventionally written (h_i) and families of planes $\{h_i\}$. Similarly individual directions are written $[u^i]$ and families $\langle u^i \rangle$.

In general 21 elastic constants are needed to define the elastic properties of crystals.[3] However this number is reduced by symmetry. Hence triclinic, monoclinic, orthorhombic, tetragonal, cubic, rhombohedral and hexagonal crystals have 21, 13, 9, 7 or 6, 3, 7 or 6, and 5 constants respectively. Isotropic materials have only 2 independent elastic constants.

Crystal defects. The basic hypothesis of crystal physics, that crystals consist of regular arrays of atoms, ions or molecules, leads to a satisfactory understanding of many physical properties such as elastic constants and specific heats. It cannot, however, explain most of the mechanical properties of crystals, particularly the notorious weakness of metals, which are some hundreds or even thousands of times weaker than expected from the perfect crystal assumption. It is thus natural to postulate the existence of crystal defects.[5-8] These can be of zero, one, two and three dimensions and may be called point, line, sheet and volume defects, respectively. All four types are known to exist and are intimately related. Thus, for example, volume defects may be considered to be clusters of point defects, sheet defects are sometimes bounded by line defects and in other cases consist of networks of line defects, and point defects are attracted or repelled by line defects.

Some of these defects are easy to visualize. Thus, voids and precipitates provide good examples of volume defects and grain and phase boundaries of sheet defects. Vacant sites in the crystal, i.e. vacancies, and intrusive atoms not located at crystal sites, i.e. interstitials, are point defects, but the line defect known as the dislocation, is not so easy to describe. The term dislocation was introduced by Love (1927) to describe a type of line defect in an elastic continuum which had been first studied by Volterra (1907) who referred to it as a distortion. The concept of a dislocation in a crystal was, however, first used by Prandtl (1921) and developed independently by Taylor, Orowan and Polanyi (1934) and Burgers (1939). The importance of the role of dislocations in the plastic deformation of crystals was well established by the early 1950s.[5-8]

Observations of crystal defects. Direct experimental evidence for the existence of dislocations and other crystal defects has been obtained since 1949 when spiral surface features were observed on crystals grown from the vapour or dilute solution.[9] These features confirmed that a crystal growth mechanism, proposed by Frank and based on a dislocation meeting the crystal surface, was operative. Later small pits were observed at the centres of these spirals after the crystals had been etched in acids. These pits were associated with dislocation lines and were observed to move when the crystals were stressed. Arrays of pits were also observed corresponding to dislocation configurations predicted by the theories. In transparent crystals three-dimensional arrays of dislocations were made visible by impurity atoms forming small precipitates on the lines. For example in silver chloride specks of silver form on the dislocations when the crystals are exposed to light.

The most powerful technique for observing crystal defects is electron microscopy which was introduced in the 1950s.[9] Thin films of crystalline material up to 100 nm thick are transparent to 100 kV electrons, but the electrons are scattered by the strain field around defects so that images can be obtained. Interpretation of the resulting micrographs is not easy as the diffraction effects may be complex but elegant theories have been developed which enable the contrast obtained from even small defects to be computed. The technique enables the motion of defects to be studied and also the many interactions which are possible between different defects. However the results are strictly only relevant to thin foils and care has to be exercised in extending the conclusions to bulk materials. This difficulty can be avoided by using the corresponding X-ray technique but unfortunately in this case no instrumental magnification is possible and long exposure times are needed. Nevertheless this method has produced valuable results for materials with low defect concentrations particularly semiconductors. More recently higher voltage electron microscopes have become available which enable thicker foils to be examined. One of the interesting and fortuitous results of using these machines is that the high voltage electrons damage the foils so that the conditions of fast nuclear reactors are simulated.

Finally the technique of field ion microscopy has enabled the atomic structure of crystal defects to be observed directly. Only high melting point materials can be used and the specimen consists of a hemispherical tip only a few hundred atoms across. Defects emerging on the surface of this tip are imaged on a screen and layers of atoms can be removed sequentially so that a three dimensional record of the structures of the defects can be obtained. Unfortunately however the specimen has to be subjected to a very large electric field so that the structures observed are rather distorted. The resolution of modern electron microscopes is also such that planes of atoms in metal crystals can be observed and hence information obtained about the structure of atomic sized defects. The experiments are however rather sophisticated and only relatively simple defects can be examined.

These experimental methods are providing a wealth of information about defects in crystals. However the results are often difficult to interpret and assess, and in particular little information has been forthcoming on the nucleation of defects. Theory therefore has an important and indeed crucial role to play in understanding the behaviour of crystal defects and in laying the foundations for utilising defect properties in designing materials for future applications.

Geometrical theories of defects

Point defects. The geometry of a single vacancy or substitutional solute atom in a crystal is well-defined, but an interstitial atom may have many possible configurations. For example a small interstitial impurity in a body centred cubic crystal such as carbon in iron may lie at one of the $\frac{1}{2}$ <100> sites which, like the equivalent $\frac{1}{2}$ <110> sites, are midway between second nearest neighbour iron atoms. Alternatively the impurity may lie at one of the larger $\frac{1}{4}$ <210> sites which are equidistant from four neighbouring iron atoms. Geometry alone can tell us little about the relative merits of these octahedral and tetrahedral sites which are illustrated in Fig. 1(a). Similarly self-interstitials may have many configurations including cases in which two atoms try to occupy

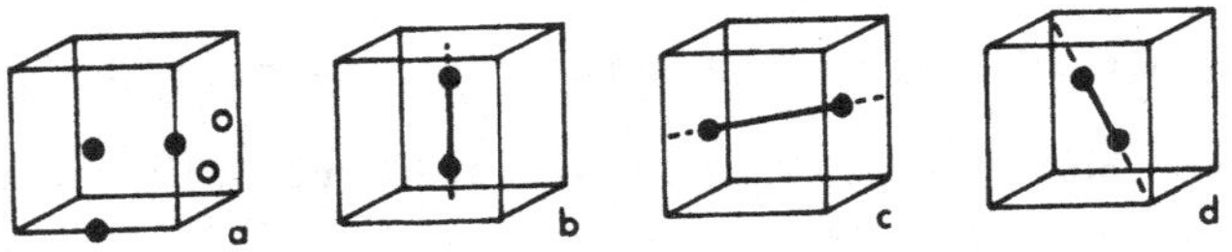

Fig. 1 Examples of (a) interstitial impurities at octahedral ½ <100> and ½ <110> sites (closed circles) and tetrahedral ¼ <210> sites (open circles) and (b)-(d) split self-interstitials along <100>, <110> and <111> respectively, in a bcc unit cell.

one lattice site. Three possible examples of these split-interstitials aligned along <001>, <011> and <111> directions in a body centred cubic crystal are illustrated in Figs. 1(b)-(d).

Of greater interest for geometrical treatment is the enumeration and classification of clusters of point defects. In particular, close packed clusters of vacancies or solutes, in which every point defect has at least one point defect in a nearest neighbour position, have been studied in detail.[10] For example the five distinct clusters of four vacancies shown in Fig. 2(a) arise for the two-dimensional square lattice. Clusters of this type consisting of n substitutional point defects may be specified by n-1 vectors each of which will be one of the group of m vectors defining nearest neighbour positions. In single lattice structures $m \leq 12$ but larger values of m occur in more complex structures, e.g. 18 vectors are required in hexagonal close packed crystals. As shown by Fig. 2(a) allowance has to be made for clusters with loop, branched and chain topology. In addition there will in general be a number of crystallographically equivalent ways of delineat-

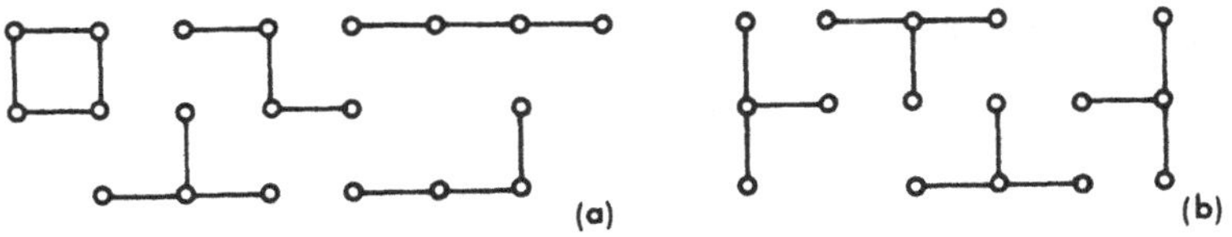

Fig. 2 Clusters of four vacancies on a two-dimensional square lattice. The five distinct clusters which are possible are shown at (a) and the four variants of the branched cluster at (b).

ing a given configuration. For example Fig. 2(b) shows the four variants of the branched cluster of Fig. 2(a). The number of variants will depend on the symmetry of both the structure and the cluster and may be as high as 48 or as low as 1. The clusters can be classified in terms of the number of bonds of different lengths linking the vacancies. This provides a measure of the binding energy of clusters. Thus the tetra-vacancy clusters of Fig. 2(e) involve six bonds. Allocating the letters A,B,C,D,E,F to the first six nearest neighbour bonds, and noting that all clusters have at least three A bonds, the remaining bonds for the five cases are given by ABB,BBC,BBD,BCD,CCF. In more complex structures distinct clusters occasionally have the same bond classification and a secondary criterion such as topology must then be invoked.[10,11]

In terms of close packed directions the three dimensional single lattice structures may be divided into 28 different groups.[10] For example, rhombohedral structures may have <100>, <$1\bar{1}0$> or [111] as their closest packed directions and therefore provide three groups. However 15 of these 28 groups have a unique close packed direction so that only linear close packed clusters parallel to this direction can arise. The remaining 13 groups can all be treated as special cases of face centred cubic, body centred orthorhombic with ½ <111> as close packed direction, and rhombohedral with <100>. Therefore an analysis of these three cases enables all possible configurations in single lattices to be examined. For example in face centred cubic there are 4 possible trivacancies, 20 tetravacancies and 131 pentavacancies, the total number of variants being 50, 475 and 4929 respectively.

The migration of clusters by means of nearest neighbour vacancy jumps can also be examined geometrically.[12] Some clusters can migrate to become different variants of themselves but in general clusters must take on different configurations and the choice may be very restricted. For example the most closely packed tetravacancy in face centred cubic crystals, which has the form of a regular tetrahedron can only migrate to one other configuration but may do so in 24 different ways. If there are c distinct configurations of n vacancies the number of ways in which cluster migration can occur may be represented by the (cxc) migration matrix M_{ij}. It can then be shown that $V_iM_{ij} = V_jM_{ji}$ where V_i is the number of variants of the cluster and no summation is implied. The matrix M_{ij} can be used to investigate repeated migration of clusters and hence equilibrium distributions.[13] Similarly the growth and contraction mechanisms of clusters can be deduced and represented by growth and contraction matrices G_{ij} and C_{ij} which satisfy $V_iG_{ij} = V_jC_{ji}$. These relations are very valuable while checking the results, which have been obtained using elementary but tedious enumeration and classification procedures. Unfortunately the problem does not lend itself very readily to computation and graph theory has not been of great assistance. Currently the work is being extended to mixed vacancy and solute clusters where migration is more difficult.[14]

Dislocations. The only important line defect in crystals is the dislocation.[5-8] Crystals deform by a process of local slip in which a small localized displacement slowly extends. The boundary in the slip plane of the slipped and unslipped regions is then a dislocation line or simply a dislocation. It is important to appreciate that the dislocation is a line defect. The process of local slip does, of course, give rise to a long-range stress field around the dislocation, but only at the line itself are atoms displaced from their normal positions to such an extent that the structure is unrecognizable.

The vector representing the magnitude of the local slip, or more generally the difference in the amounts of local slip in neighbouring regions, is known as the Burgers vector. It is associated with the whole dislocation line and is thus constant for the line. It follows that dislocations must end at either the surface of a crystal or at another dislocation. Where two or more dislocations meet, one has a dislocation node. The sum of the Burgers vectors on either side of such a node must be equal. This rule is subject, however, to using a consistent sign convention for the direction of the Burgers vector.[15,16]

The atomic arrangements near a dislocation line will depend on the relative orientations of the line $\underset{\sim}{\ell}$ and the Burgers vector $\underset{\sim}{b}$. Edge and screw dislocations correspond to $\underset{\sim}{b}$ being perpendicular and parallel to $\underset{\sim}{\ell}$

respectively. Edge dislocations have an extra half-plane of atoms which may be above or below the slip plane defining positive and negative types. Screw dislocations have no extra half-plane but may be right or left handed. By definition they must be straight whereas edge dislocations may be curved. In general dislocations have mixed character being partly edge and partly screw.

In order to define the Burgers vector of a dislocation uniquely use is made of the Burgers circuit. This is a closed right handed (RH) circuit of steps delineated in good crystal around a dislocation. The same circuit is then repeated in a perfect reference lattice where it will not close. The closure failure from the finish F to the start S of this circuit defines the Burgers vector $\underset{\sim}{b}$ of the dislocation. This is illustrated in Fig. 3 for both edge and screw dislocations in a simple cubic crystal. Note that the RH/FS convention adopted here defines the Burgers vector in the reference crystal. Other conventions use left-handed circuits, equate $\underset{\sim}{b}$ to $\underset{\sim}{S}\underset{\sim}{F}$ or define the Burgers vector in the distorted real crystal by closing the circuit in the reference crystal. Each of these changes reverses the sense of the Burgers vector.

The dislocations shown schematically in Fig. 3 have Burgers vectors which are the shortest possible lattice vectors of the simple cubic structure. In real materials the situation is more complex. For example in body centred cubic crystals $\underset{\sim}{b}$ is the shortest lattice vector $\frac{1}{2}$ [111] of magnitude $\sqrt{3}/2$. However the interplanar spacing of the (111) planes which are perpendicular to this direction is $\sqrt{3}/6$. Therefore the extra $\frac{1}{2}$-plane of the edge dislocation in these crystals consists of three

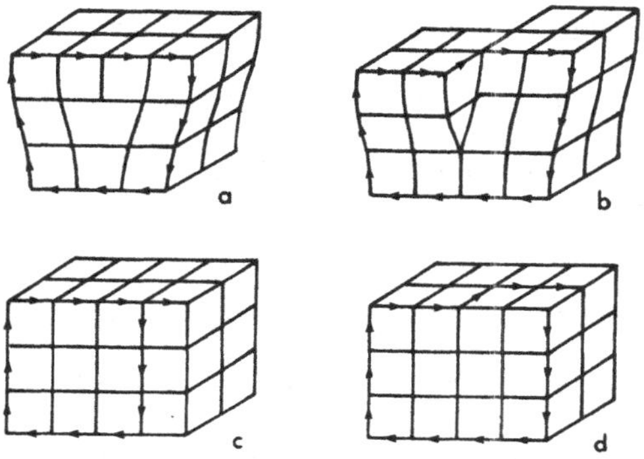

Fig. 3 Burgers circuits in real crystals containing (a) an edge dislocation and (b) a screw dislocation and repeated in reference crystals (c) and (d) respectively to define the Burgers vectors FS.

sheets of atoms. In practice these three extra ½-planes do not remain in contact but separate in the slip plane. Similarly dislocations in face centred cubic crystals with ½ [110] Burgers vectors dissociate into two partial dislocations with Burgers vectors $\frac{1}{6}$ [211] and $\frac{1}{6}$ [12$\bar{1}$]. It is not possible however to predict the extent of these dissociations using geometrical models.

Interfaces. Crystalline interfaces may be coherent, semi-coherent or incoherent.[17-19] Fully coherent boundaries arise in faulting, in twinning and in a few martensitic phase transformations. They involve perfect matching of the crystal structures on either side of the interface. Semi-coherent boundaries include low-angle grain boundaries and many martensitic interfaces and may be glissile or non-glissile in character. Their structure can be described formally in terms of an array of dislocations. Incoherent boundaries are those in which the formal dislocation content is so high that individual dislocations have no physical meaning. However some geometrical models of high angle grain boundaries have been developed. Examples of the three types of boundary are shown schematically in Fig. 4.

In order to discuss interfaces, it is necessary to develop a formalism for describing the relationship between the lattice points of two crystals which have different structures and/or orientations.[17-19] The mathematical treatment of this problem, is essentially that of finding an affine transformation, i.e. a real or hypothetical finite homogeneous deformation, which will carry one set of lattice points into another. The concept of a fully coherent interface then requires that one plane be invariant during this deformation. The conditions for this are rather restrictive, but they are readily satisfied for an enumerable infinity of cases when the two lattices differ only in orientation. If two lattices in some particular orientation relationship have a matching plane in common, it is always possible to find a second matching plane corresponding to a different orientation relation between the same two lattices; the two planes are generally not crystallographically equivalent, and need not be rational. This point is illustrated schematically in Fig. 5 which shows the homogeneous shear of a sphere into an ellipsoid. The two bodies intersect in two circles which define the two undistorted planes of the deformation. The first is the plane on which the shear

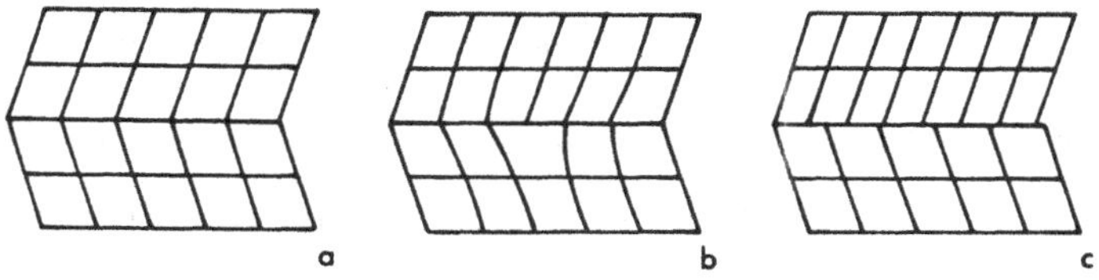

Fig. 4 Schematic examples of (a) coherent, (b) semi-coherent and (c) incoherent interfaces.

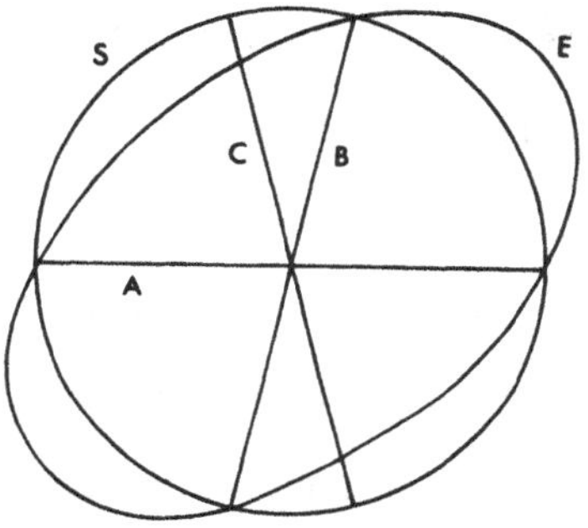

Fig. 5 Homogeneous shear of a sphere S into an ellipsoid E to produce two undistorted planes A and B. Plane A is invariant and B becomes invariant after being rotated to its original orientation C.

occurs and is invariant. The second is not invariant but may be made so by rotating it to its original orientation. These two invariant planes define matching interfaces between a lattice embedded in the sphere and a lattice in the ellipsoid.

When a coherent interface is geometrically impossible, the misfit at a planar interface may be described formally in terms of dislocations.[18] The general expression for the net Burgers vector of the dislocation lines crossing any unit vector in the interface, and the concept of the surface dislocation density tensor may be introduced. The multiplicity of possible descriptions of the lattice relations leads to corresponding variations in the calculation of the net Burgers vectors of the interface dislocations. When the density of Burgers vector is sufficiently small, the interface misfit becomes concentrated into linear regions of discontinuity which may appropriately be described as interface dislocations. Both the formal dislocation description of an interface and the more physical model of an interface containing discrete dislocation lines may be treated in a slightly different way by means of the theory of the "O-lattice".[20]

Theories of the crystallography of deformation twinning attempt to predict the shear plane and direction and the magnitude of the shear strain which will be operative in any particular crystal.[4,21,22] In many cases the twinning mode with the smallest possible strain, consistent with restoring the lattice in a new orientation, is operative. However in multiple lattice structures shuffling of groups of atoms is in general necessary so that the complexity of the shuffling process has also to be taken into account.[23-26] Fig. 6 shows schematically the type of atomic shuffling which is necessary in crystals like the hexagonal close packed metals and materials with the diamond structure which have two atoms at each lattice point.

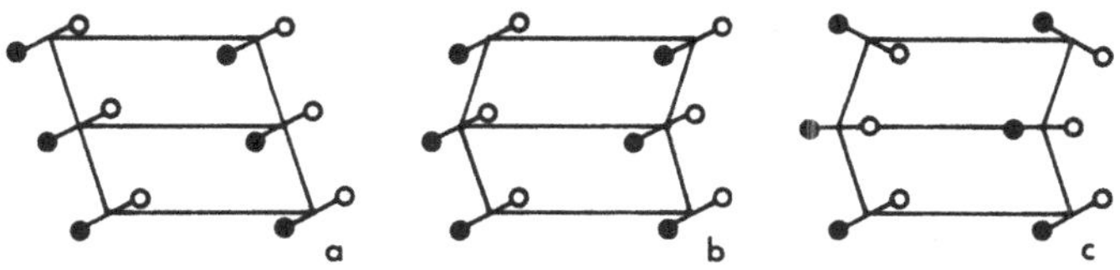

Fig. 6 A possible atomic shuffling mechanism associated with twinning of a lattice with two atoms at each point. The single crystal (a) shears to become the structure at (b) which requires shuffles to generate the twin at (c). The open and closed symbols lie in front of and behind the paper respectively.

Martensitic transformations involve a change of crystal structure.[17-1] The parent and product structures meet at a plane interface but the homogeneous deformation responsible for the phase change simply leaves a cone of undistorted lines and no invariant plane. Therefore an additional lattice invariant deformation slip, twinning or faulting, is necessary. This second deformation shears the cone of undistorted lines by a critical amount so that an undistorted plane is obtained. This can then be rotated to its original orientation to become invariant. The following simple matrix algebra formulation of this process has been developed.[19,27] Let the invariant plane strain of the transformation be

$$\underset{\sim}{F} = \underset{\sim}{I} + f\, \underset{\sim}{u}\, \underset{\sim}{h}^T$$

where $\underset{\sim}{I}$ is the unit matrix and f, $\underset{\sim}{u}$ and $\underset{\sim}{h}$ are the magnitude, direction and plane normal of the strain. Also $\underset{\sim}{F} = \underset{\sim}{D}\, \underset{\sim}{S}$ where $\underset{\sim}{D}$ is the lattice deformation and $\underset{\sim}{S}$ the lattice invariant deformation. Then $\underset{\sim}{D} = \underset{\sim}{R}\, \underset{\sim}{P}$ where

$\underset{\sim}{R}$ is a rotation and $\underset{\sim}{P}$ a pure strain so that $\underset{\sim}{F} = \underset{\sim}{R}\,\underset{\sim}{P}\,\underset{\sim}{S}$. Let $\underset{\sim}{P}\,\underset{\sim}{S} = \underset{\sim}{A}$, known as the characteristic strain and using $\underset{\sim}{R}^T\underset{\sim}{R} = \underset{\sim}{I}$ eliminate $\underset{\sim}{R}$ to obtain $\underset{\sim}{F}^T\underset{\sim}{F} = \underset{\sim}{A}^T\underset{\sim}{A}$, $= \underset{\sim}{Q}$ say, so that $\underset{\sim}{X} = \underset{\sim}{F}^T\underset{\sim}{F} - \underset{\sim}{Q} = 0$. Finally substituting for $\underset{\sim}{F}$ gives

$$X_{ij} = f^2h_ih_j + f(u_ih_j + u_jh_i) + \delta_{ij} - Q_{ij} = 0 \ .$$

As X_{ij} is symmetric this provides six equations for the six unknowns f, h_1/h_3, h_2/h_3, u_1/u_3, u_2/u_3 and g, where g is the magnitude of $\underset{\sim}{S}$. Thus by choosing a correspondence between the two structures and hence the pure strain $\underset{\sim}{P}$, and a plane and direction for the lattice invariant shear $\underset{\sim}{S}$, the invariant plane strain $\underset{\sim}{F}$ can be obtained. The analysis has been applied successfully to many transformations but in some cases more elaborate mechanisms are involved.[17,19,27]

Grain boundaries are often found to have, at least approximately, special orientations in which a fraction of the lattice points of the two grains coincide. These coincidence site lattice or CSL boundaries are characterised primarily by Σ, the reciprocal of the fraction of sites which coincide.[20] Twin boundaries in body centred and face centred cubic crystals have $\Sigma = 3$ and small values of Σ are usually thought to be associated with low energies. However boundaries with $\Sigma > 100$ have also been reported as having significance.

Continuum theory of defects

Isotropic elasticity theory has played a central role in the development of theories of crystal defects particularly dislocations. Indeed the word dislocation was first used by Love in a treatise on elasticity. Other defects tend to have short-range strain fields and therefore their properties are intimately related to the detailed atomic structure of their core regions. However dislocations have long range strain fields and therefore many of their properties can be adequately described using analyses based on elasticity.

A dislocation can be introduced very simply into an elastic medium by first removing a cylinder of material of small radius δ along the line which is to be the site of the defect.[5-8] The material is then cut up to this line and the two sides of the cut displaced relative to each other by the Burgers vector. The two parts of the material are then welded together and allowed to relax. The stress field in the medium then approximates to that around a dislocation, infinite stresses along the defect being avoided through the presence of the hole. For screw and edge dislocations the displacement of the faces of the cut must be parallel and perpendicular to the hole respectively. These cases are illustrated in Fig. 7.

Solutions of the equations of elasticity for dislocations now give a stress-field proportional to $\mu b r^{-1}$ where μ is the shear modulus, b the magnitude of the Burgers vector and r is the distance from the core. For edge dislocations there is angular dependence but not for screws. The elastic energies of dislocations can be calculated from the stress-field

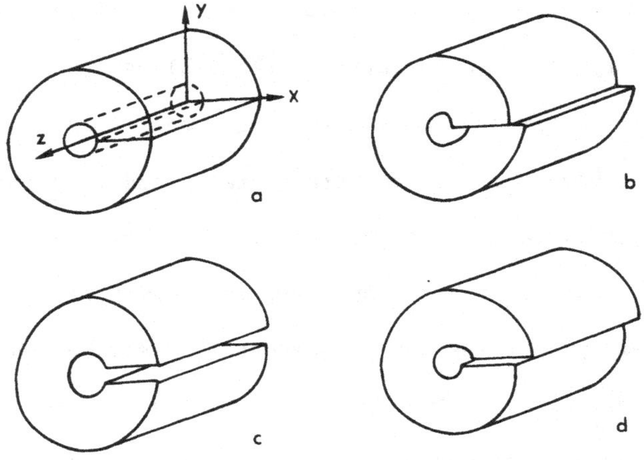

Fig. 7 Elastic models of dislocations. The cylinder at (a) is cut on the z-x plane up to the cylindrical hole along the z-axis. Relative displacements of the two faces of the cut parallel to (b) the x-axis and (c) the y-axis now generate edge dislocations, and to (d) the z-axis screw dislocations.

and are proportional to $\mu b^2 \ell n(R/\delta)$ where R is the outer radius of the elastic block. In real situations this energy is of the order of $5 \times 10^{-9} Jm^{-1}$. The energy of the core region, which is excluded here, can be shown to be much less than this. The force per unit length experienced by a dislocation due to an applied shear stress p is F = pb and this may cause the dislocation to become curved with a radius of curvature R given by $p \approx \mu b R^{-1}$. There is an elastic interaction between dislocations so that for example two like parallel screw dislocations a distance r apart repel each other with a force $\mu b^2/2\pi r$. As the elastic strain energy of a dislocation decreases as the defect approaches a free surface, it is attracted to the surface. This may be described in terms of an image force acting on the dislocation. Of crucial importance is the fact that the energy of a dislocation is proportional to b^2. This means that a dislocation with Burgers vector 2a, where a is a lattice vector, will always dissociate into two dislocations of Burgers vector a, thus reducing the combined energies of the defects by a factor 2.

The treatment of the elastic properties of a straight dislocation using anisotropic elasticity theory involves finding the roots of a sextic equation the coefficients of which are functions of the elastic constants.[7] Thus except in a few special cases in high symmetry crystals, such as dislocations parallel to <110> or <111> in cubic materials, solutions must be obtained numerically. In many cases the resulting stress fields and energies do not differ appreciably from the isotropic results but sometimes the use of anisotropy is crucial. For example crystalline mercury, which freezes at - 40°C, is very anisotropic, Young's modulus and the shear modulus varying by factors of 7 and 11 for different directions and shear systems.[28] It is therefore not surprising as shown in Fig. 8 that the elastic energies of the dislocations in this metal are also very anisotropic.[29] Indeed the easiest slip system is not the one with the smallest Burgers vector. Another important application of anisotropic elasticity is to grain boundaries which cannot exist in isotropic media.[30]

Elementary applications of elasticity theory to dislocations concern either straight dislocations or circular loops. In practice dislocations in crystals take on irregular forms and it is thus necessary to develop analyses which can be applied to other configurations. For example curved dislocations can be treated as series of dislocation segments and

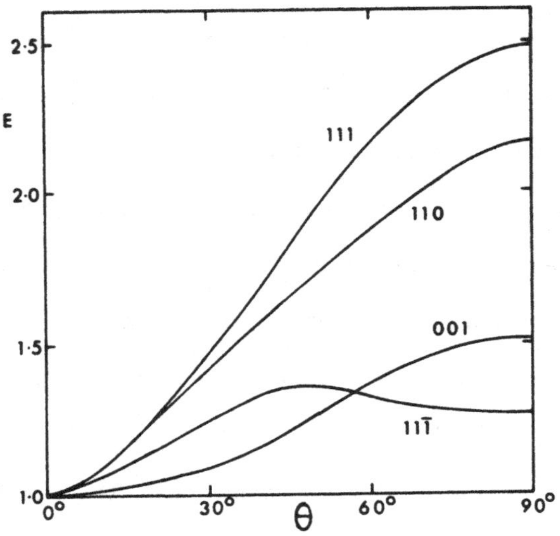

Fig. 8 Anisotropic elastic energies E of dislocations with Burgers vector $\underset{\sim}{b} = \frac{1}{2} <1\bar{1}0>$ lying on (111), {110}, {001} and {11$\bar{1}$} planes in crystalline mercury. The orientation of the dislocation is given by the angle θ where θ = 0 and π/2 correspond to screw and edge dislocations respectively. Using isotropic elasticity $E(\pi/2)/E(0) = 1.5$.

equilibrium configurations determined by minimising the total energy. Powerful analytical techniques including Green's function and integral formalisms have been developed to handle these situations but in general recourse has usually to be made to extensive computation.[31]

The limitations of elasticity theory when applied to dislocations are illustrated by the case of a jog or step on an otherwise straight dislocation line.[32] There are three cases to consider in which the Burgers vector is (Z) parallel to the dislocation line, (Y) parallel to the jog, and (X) perpendicular to both line and jog. The results are shown in Fig. 9, for all three cases, as a function of jog length. The basic application of the theory shows that short jogs, and even jogs of zero

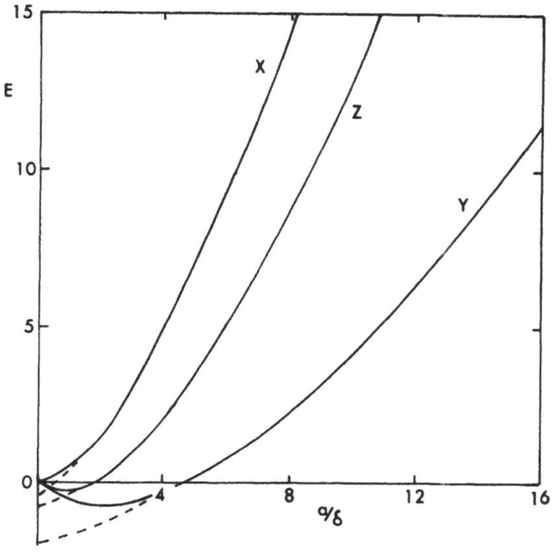

Fig. 9 Isotropic elastic energies E of X, Y and Z type jogs of length a on a straight dislocation of core radius δ. The broken curves show errors which arise when the work done by core surface tractions is not included.

length, have negative energies. However if the work done by the core surface tractions when the defect is generated is included the anomalous energies for zero-length jogs are avoided but short Y and Z jogs still have negative values. Thus the theory can not be used for the important case of short jogs.

Point defects can be handled by elasticity through considering a spherical inclusion of radius $(1 + \varepsilon)r_0$ in a hole of radius r_0.[33] The relaxed radius of the inclusion will be $(1 + \alpha\varepsilon)r_0$ and the displacement field is then

$$\underset{\sim}{u}(\underset{\sim}{r}) = \alpha\varepsilon r_0{}^3 r^{-3}\underset{\sim}{r}$$

where $\alpha = (1 + \nu)[3(1 - \nu)]^{-1}$. The volume change is $\Delta V = 4\pi r_0{}^3\varepsilon$ and the formation energy is $E^F = 4\pi r_0{}^2\gamma - 2\pi\alpha\gamma^2 r_0\mu^{-1}$, where γ is the surface energy. Various interaction energies can also be obtained, but the analysis is unable to provide values for ε and γ and cannot cope with more complex point defects and clusters.

It is interesting to note that continuum theories of defects also owe a great deal to hydrodynamics and electro-magnetic theories.[7,8] Analogies between the different physical phenomena have enabled striking developments to be made. There have also been considerable efforts to develop continuous distribution theories for defects. For example in the case of dislocations the density of the defects is allowed to tend to infinity while the Burgers vector of the individual dislocations tends to zero, leaving the resulting total Burgers vector constant.[34,35] Finally extensions of continuum models to incorporate in part the discrete nature of materials have also been attempted.[36,37]

Discrete theories of defects

Inter-atomic potentials and computer models. Geometrical theories provide only a general framework in terms of which the detailed properties of crystal defects may be investigated. Continuum theories give this detailed information for long range effects but have difficulty in providing adequate information for the important short range phenomena such as nucleation, combination and dissociation of defects. To investigate these intimate properties of defects the discrete nature of the structure of the core regions of the defects needs to be taken into account. A successful method of doing this is to use inter-atomic potentials to determine the structure, and self-, interaction- and migration-energies. However because of the large number of atomic interactions involved the method relies heavily on computational techniques. Two approaches are possible using either direct- or reciprocal-lattice procedures.[33] The latter method is more elegant but is restricted to relatively simple short range potentials and to fairly elementary defects. On the other hand the direct lattice method relies more heavily on computing, but can in principle be used for any potential and for complex defects. This is the method to be discussed here.

Ideally it is desirable to use inter-atomic potentials derived from first principles in these computations. However unfortunately such potentials often have difficulty in predicting the correct crystal structure and can not therefore be expected to give reliable information on the structure of crystal defects. Similarly pseudo-potentials have severe limitations in describing defect structures. Thus in computer simulation studies, empirical 2-body central potentials are often used. These may consist of several piecewise continuuous cubic splines. At the knots, where the splines meet, the first and second derivatives are normally made continuuous and the potential terminates at an appropriate inter-atomic separation at zero slope. For example the iron and copper potentials of Fig. 10, which are used widely, terminate between second

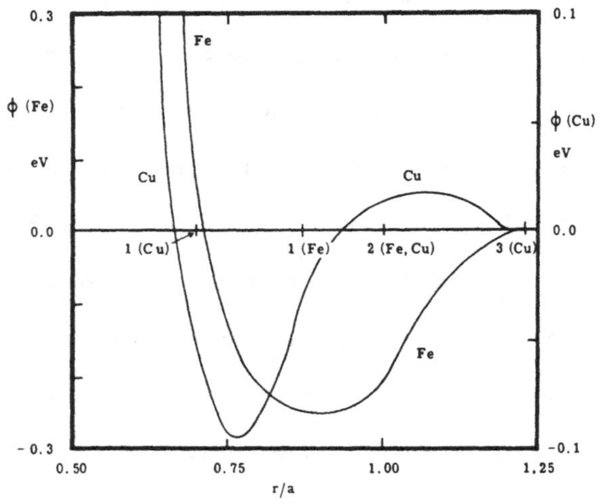

Fig. 10 Interatomic potentials $\phi(r)$ for iron and copper with different energy scales marked at the left and right respectively. Nearest neighbour distances are indicated on the r/a axis, where a is the lattice parameter.

and third neighbours.[38-40] Ideally longer range potentials could be used, but this would increase the computation time appreciably.

Empirical potentials are matched to as much reliable experimental information as is available. This normally includes, the lattice parameters, the elastic constants, critical information from the phonon dispersion curves, stacking fault energies and vacancy formation and migration energies. In practice most applications have been to cubic crystals so that only one lattice parameter and three elastic constants are involved. However even in this relatively simple situation it is not in general possible to match the elastic constants using a pair potential which holds the crystal in equilibrium at the correct lattice parameter.

One method of overcoming this problem is to introduce a second volume dependent component to the potential which may be interpreted as the pressure associated with the free electron gas. This can however give rise to problems in interpreting the formation volumes and free energies of defects.[40]

The spline potentials $\phi(r)$ for b.c.c. iron[38] and f.c.c. copper[39,40] shown in Fig.10 are defined by

$$\phi(r) = \phi_i(r) , \qquad r_i < r < r_{i+1}$$

$$\phi_i(r) = \sum_{j=0}^{3} A_{ij}(r - r_i)^j , \qquad i = 1, 2, 3 \ldots..$$

where the knots are at $r = r_i$ and A_{ij} are the coefficients of the cubic splines. The iron potential has three splines and the copper potential nine. For b.c.c. crystals the elastic constants c_{11}, c_{12} and c_{14} are matched to the potential using the relations

$$3ac_{11} = \phi''(r_A) + 2r_1^{-1}\phi'(r_A) + 2\phi''(r_B)$$

$$3ac_{12} = \phi''(r_A) - 4r_1^{-1}\phi'(r_A) - 2r_2^{-1}\phi'(r_B)$$

$$3ac_{44} = \phi''(r_A) + 2r_1^{-1}\phi'(r_A) + 2r_2^{-1}\phi'(r_B) ,$$

where $\phi'(r)$ and $\phi''(r)$ indicate first and second derivatives and r_A and r_B the distances of first and second nearest neighbours. The vacancy formation energy E_V^F is given by $E_V^F = - 4\phi(r_A) - 3\phi(r_B) - E_V^R + p\Omega_V^F$, where E_V^R is the relaxation energy, Ω_V^F the formation volume, and p the Cauchy pressure. Similar expressions arise for f.c.c. crystals. The iron potential of Fig. 10 holds the model crystal in equilibrium at the correct lattice parameter so that in this case p = 0.

The procedures used in deducing the structures and energies of defects [41] now involve constructing a computer model of a perfect crystallite. Large computers enable up to about 10^4 discrete atomic sites to be included. A first approximation of the defect to be studied based on simple geometrical or elastic models is then introduced. The atoms are then allowed to relax to their minimum energy positions using an interatomic potential. This effectively involves calculating the net force on every atom and hence, using Newton's equations of motion, deducing the configuration of the atoms after an increment of time. This process is then repeated until the equilibrium configuration is reached. In practice however powerful numerical procedures such as the method of conjugate gradients,[42] are used in the relaxation routine.

The boundary conditions used in this process are of course of crucial importance.[41] In many cases periodic boundaries may be used realistically so that the model is effectively infinite in extent. Alternatively it may be adequate to use rigid boundaries, allowing any volume changes to be made between different relaxation sequences. In some cases it may be necessary to use flexible elastic boundaries which may relax to take up changes of shape within the computational cell. In all cases a mantle of atoms must effectively surround the relaxing crystallite so that atoms at or near the boundary have a full complement of neighbours.

Using these methods care has to be taken to ensure that the initial configuration is a reasonable approximation to the correct structure. Otherwise the model might explode, particularly if atoms are much closer together than nearest neighbour separation, so that there is a large repulsive force between them. Also care has to be exercised to avoid local energy minima and metastable structures. In practice however it is often found that the resulting structures, although at first surprising, are plausible in terms of acceptable packing arrangements of the atoms. The main limitations of the method are (a) the reliability of the potentials used, (b) the small size of the model, (c) the restriction to equilibrium structures implying zero temperatures, (d) the lack of generality in the results and (e) the difficulty of checking the results.

Point defects. In principle it is very easy to study a single vacancy using computer simulation techniques.[33] An atom is simply removed from the interior of the model and the surrounding atoms allowed to relax to their new minimum energy locations. However, in practice, although the displacements of these atoms are small they are found to be very anisotropic. For example in f.c.c. crystals the first nearest neighbours relax inwards but the second neighbours outwards, and many of the displacements are not radial. Problems therefore arise in selecting appropriate boundary conditions. Again when calculating the formation energy of a vacancy allowance has to be made for the fact that the atom removed must be placed on the surface of the model so that effectively on average only about one-half of its bonds are broken. For non-equilibrium potentials there is also the problem of calculating the volume change associated with the defect, which is used in determining the volume dependent part of the energy. As the vacancy formation energy is part of the experimental data to which the potential is matched it is crucial that this term be calculated accurately, but different procedures unfortunately have tended to give very different answers. Despite these problems reliable computations have now been carried out for vacancies in several different metals.[40]

Simulations of point defects based on determining relative rather than absolute energies do not suffer these problems. For example divacancy binding energies E_B^{2V} have been examined in iron.[43] These energies are given by

$$E_B^{2V} = E_F^{2V} - 2E_F^V$$

where $E_F^V = E^{P+V} - E^P$ and $E_B^{2V} = E_F^{2V} - 2E_F^V$. Here E^P, E^{P+V} and E^{P+2V} are the energies of the model crystal when perfect, when containing a single vacancy and when containing a divacancy. The values of E_B^{2V} for first $\frac{1}{2}$ <111>, second <100> and fourth $\frac{1}{2}$ <113> nearest neighbours are found to be 0.13 eV, 0.19 eV and 0.05 eV respectively, the third <110> neighbour divacancy being unstable. It is interesting to find that the most stable divacancy is the second and not the first nearest neighbour. The variation of E_B^V as a function of applied stress has been determined and found to vary by up to 50% for strains of 3%.[43]

The migration of vacancies can readily be simulated by placing a self-interstitial at different locations along the axis of a divacancy and plotting the resulting energy barrier.[44] For the case of iron this barrier is shown in Fig. 11. It has two symmetrically located maxima, corresponding to the two (111) planes through which the migration self-interstitial must pass in moving from its initial atomic site at [000] to a vacant site at $\frac{1}{2}$ [111]. The height of the barrier, or migration energy E_M^V, is 0.68 eV. As indicated in Fig. 11 this is increased to 0.75 eV by a 3% tensile strain parallel to [111] but decreased to 0.58 eV by a compressive strain.[44] These changes may be readily interpreted in terms of the compensating contraction and expansion of the (111) planes through which the migrating atom passes. Migration of divacancies in b.c.c. crystals is interesting as the favoured <110> configuration cannot migrate to a variant of itself by means of a single nearest neighbour vacancy jump. It may become of $\frac{1}{2}$ <111> type for which the migration energy E_M^{2V} = 0.78 eV or of $\frac{1}{2}$ <311> type with E_M^{2V} = 0.66 eV. The energies of the reverse jumps are 0.72 eV and 0.52 eV respectively, so that surprisingly the mechanism involving the fourth nearest neighbour $\frac{1}{2}$ <311> divacancy is preferred.[44]

Larger clusters of vacancies have been studied in f.c.c. crystals. In particular the 4 trivacancies and 20 tetravacancies in copper have been considered in detail and have provided fascinating results.[40] For example the most favoured trivacancy is the equilateral triangle which collapses to become a tetrahedron of vacancies surrounding a central interstitial. Several tetravacancies including the rhombus shaped cluster on {111} and the square on {001} also collapse. The former provides the nucleus for a faulted dislocation loop and the detailed structure suggests a mechanism for cross-slip on to a different variant

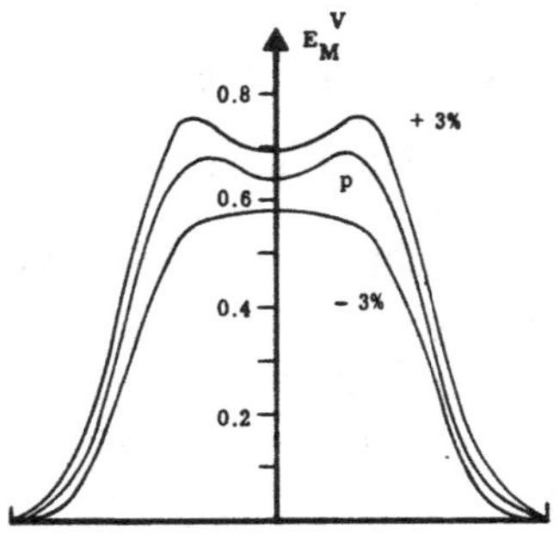

Fig. 11 Potential energy barriers for migration of a vacancy in iron. The curves are for an unstrained crystal p and for crystals strained by ± 3% in the [111] direction which is parallel to the migration path.

of {111}. The square collapses to an octahedron containing a di-interstitial. If these interstitials are eliminated by the arrival of two more vacancies, a further 8 atoms collapse into the cluster to provide a 14 vacancy plus 8 interstitial configuration. Remarkably removal of the interstitials then leaves a stable cubic 14-vacancy cavity. This mechanism, which is illustrated in Fig. 12, would thus appear to provide a nucleation process for voids in f.c.c. metals.[40,45]

Dislocations. The simulation of dislocations is carried out by calculating the elastic strain field of the defect, fixing the atoms in the mantle at these locations and then allowing the atoms in the computational cell to relax.[46-48] Normally only straight dislocations are considered and periodic boundary conditions can be applied on the faces of the model perpendicular to the line. Because of the long range strain

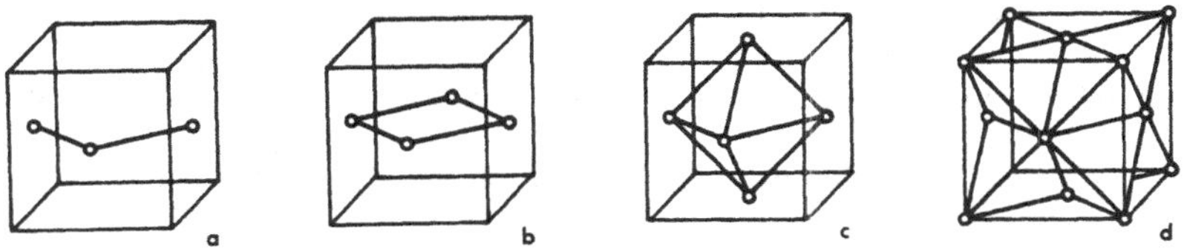

Fig. 12 Growth of a 14-vacancy void in a f.c.c. crystal. The right-angled trivacancy (a), grows to become the square tetravacancy (b) and hence the octahedral hexavacancy (c). Removal of eight further vacancies at the corners of the cube then generates the tetrakaidecavacancy (d) which is a stable void in the form of a stellated octahedron.

fields of dislocations, models of large dimensions are needed perpendicular to the line. This is particularly true of the slip plane as the dislocations tend to dissociate in this plane.

The edge dislocation with Burgers vector $\frac{1}{2}$ [111] lying in the $(1\bar{1}0)$ slip plane of b.c.c. crystals has been examined in detail using several different potentials.[47] This dislocation lies along the $[11\bar{2}]$ direction which is not perpendicular to a mirror plane. Atomic displacements are therefore expected both parallel and perpendicular to the dislocation. This is predicted by anisotropic elasticity theory but not isotropic theory. However the discrete models suggest that in practice the displacements in the core region are much larger than those prediced by elasticity. In particular the dislocation core is much wider in the slip plane. One of the difficulties with this method is designing a convenient procedure for displaying the results. The simulation provides the relaxed locations of the atoms and hence their displacements but plots of these displacements can be very confusing especially when they are three dimensional. However the relaxed structure of the $\frac{1}{2}$ [111] (1$\bar{1}$0) edge dislocation in iron shown in Fig. 13 clearly shows the three extra half-planes associated with the defect when projected on to the (11$\bar{2}$) plane.[48] Note that in Fig. 13(a) no attempt has been made to distinguish between atomic sites lying in the six (11$\bar{2}$) planes which form the stacking sequence. Thus although (111) is not a mirror plane the 2-fold axis perpendicular to (1$\bar{1}$0) makes the projection appear symmetric.

The most important property of a dislocation is the stress required to make it move. In an otherwise perfect crystal this stress is often small and this is illustrated clearly in the computer models. Indeed it is found that when a vacancy is placed near a dislocation, the dislocation may glide towards the vacancy.[47] However care has again to be exercised in deducing quantitative information from the simulations as the fixed boundary conditions are based only on the initial location of the dislocation.

Interfaces. Computer simulation methods are particularly well-suited to investigating the structure of planar crystalline interfaces, as the problems are essentially one-dimensional. In particular twin, tilt and twist boundaries have been studied in detail. The {112} twin boundary in b.c.c. metals proved particularly interesting.[39 41] It is normally assumed that the classical orientation relation of reflection in the interface is satisfied for this twin at both a macroscopic and at an atomic level. However simulation studies have demonstrated that an alternative structure involving an additional translation of $\frac{1}{12}$ <11$\bar{1}$> parallel to the interface may sometimes occur. The two structures are

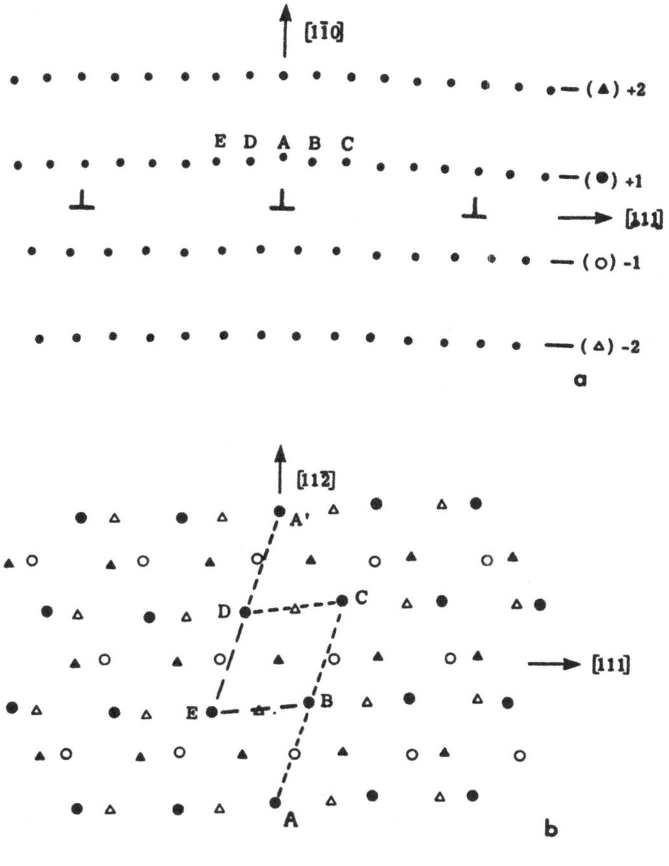

Fig. 13 Structure of a ½ [111] (1$\bar{1}$0) edge dislocation in iron projected on to (a) the (11$\bar{2}$) plane and (b) the (1$\bar{1}$0) plane. The different symbols in (b) correspond to the four (1$\bar{1}$0) planes labelled ± 2, ± 1 in (a), where the six (11$\bar{2}$) planes comprising the stacking sequence are shown superimposed. The three extra half-planes associated with the dislocation are indicated by conventional symbols in (a). Pipe diffusion of vacancies occurs along the equivalent paths ABCDA' and ABEDA'.

shown in Fig. 14. For models representing iron the two boundaries were found to have similar energies so that in practice both should be found. The volume increase associated with this interface can be deduced by moving the outer boundaries of the model until long range strains are eliminated.

The structures of several tilt boundaries in the f.c.c. metal copper have also been examined and the results illustrate several interesting features.[49] The {111} tilt (or twin) boundary has the lowest energy and exhibits little relaxation but the {113}, {120} and {112} boundaries all involve translations away from the conventional mirror image orientation relation. In addition for some boundaries several planes of atoms near the interface coalesce, at least partially, so that the boundaries may be broad and asymmetric. This is illustrated in Fig. 15 for the {112} tilt boundary and is again contrary to what is normally assumed. Twist boundaries in copper and nickel have also been examined.[39,50,51] The ones chosen have been high angle co-incidence site lattice boundaries on {001} planes with Σ = 5, 13, 17 and 25 and on {110} planes with Σ = 3 and 11 and again translations parallel to the interface are found to occur. For {001} boundaries as shown in Fig. 16 the translations leave the structures highly symmetric but {110} twist boundaries are deduced to have low symmetry. The role of symmetry in defining these structures is therefore of considerable interest.

The significance of in-plane translations in these relaxed structures is that several crystallographically equivalent but distinct structures may arise. In practice if two adjacent regions of boundary have different

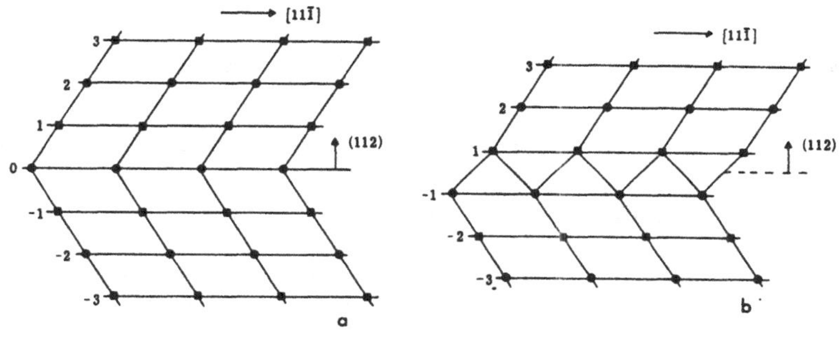

Fig. 14 Two structures for a (112) twin boundary in b.c.c. crystals shown projected on to the (1$\bar{1}$0) plane. Atoms represented by closed circles and squares lie on adjacent (1$\bar{1}$0) planes. The conventional reflection twin is shown at (a) and the alternative isosceles twin at (b).

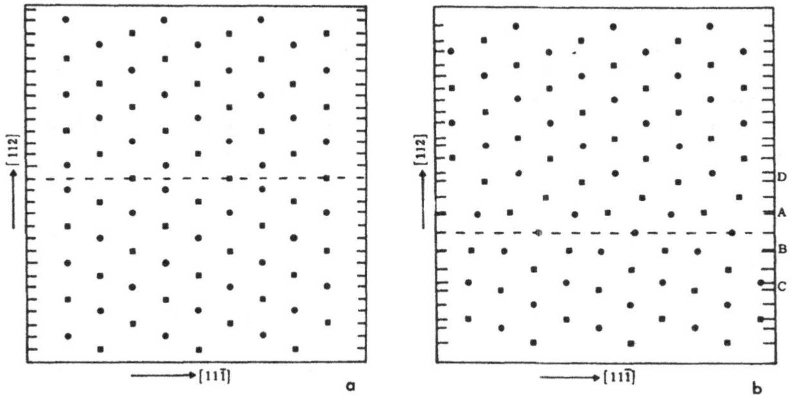

Fig. 15 The (112) tilt boundary in copper projected on to (1$\bar{1}$0). The conventional unrelaxed mirror-image structure is shown at (a) and the relaxed structure at (b), circular and square symbols representing atoms on adjacent (1$\bar{1}$0) planes. A translation of $\frac{1}{4}$ [1$\bar{1}$0] + 0.1623 [11$\bar{1}$] relates (b) to (a) and further relaxation results in the pairs of planes marked A, B, C, D in (b) being at least partially coalesced.

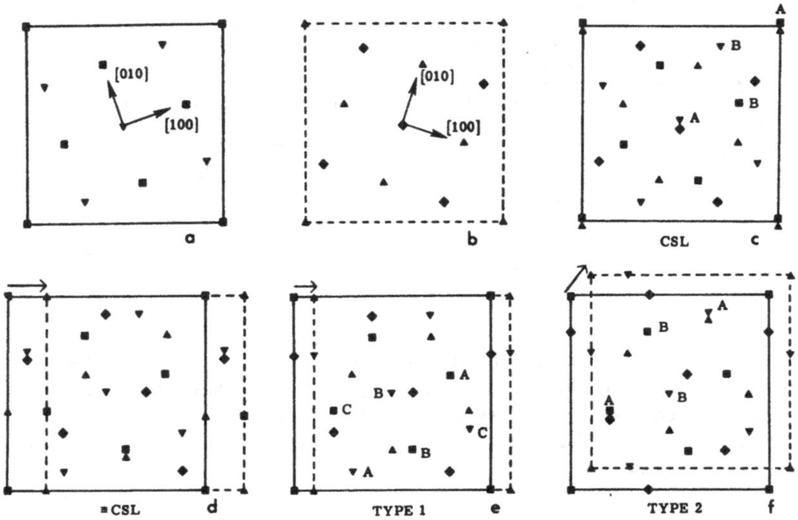

Fig. 16 Structures of possible $\Sigma = 5$ CSL (001) twist boundaries in f.c.c. crystals. The two grains represented by cells shown by continuous and broken lines in (a) and (b) are superimposed to give the CSL structure (c). In-plane translations given by the arrows in (d), (e) and (f) result in an equivalent CSL structure and two types of alternative high symmetry structures respectively. Triangular and square symbols represent atoms in odd and even planes as counted from the interface. Double symbols in (c), (d) and (f) represent superimposed sites.

in-plane translations they will be separated by a grain boundary dislocation with a Burgers vector characteristic of the particular boundary. Burgers vectors of this type will normally be much smaller than those of crystal dislocations and hence have low energies. They may thus occur in high densities. In addition an array of these dislocations will cause an additional rotation at the interface, either tilt, twist or mixed in character. Also such dislocations may be seen using transmission electron microscopy techniques providing a convenient check on predictions based on computer simulation models.

Interactions between defects. Interactions between defects of different types can also be investigated advantageously using discrete models. This may be conveniently illustrated using the case of vacancy/twin boundary interactions in b.c.c. metals.[52] The energies of the perfect crystal E^P, the crystal with a twin E^{P+T}, the crystal with a vacancy E^{P+V} and of the crystal with a twin and a vacancy E^{P+V+T} are first calculated. The energies of the twin boundary $E^T = E^{P+T}-E^P$ and of the vacancy $E^V = = E^{P+V}-E^P$ are then deduced, and finally the interaction energy $E_I^{TV} = E^{P+T+V}-E^P-E^T-E^V$ of the twin and the vacancy. These interaction energies are found to be negative when the vacancy is near the twin so that there is a positive binding energy. However, rather surprisingly, maximum binding occurs when the vacancy is adjacent to, but not in the boundary. This is true for both types of {112} twin boundary. The magnitudes of these binding energies are of the order of 15% of the vacancy formation energies.

The migration of vacancies along twin boundaries has also been simulated. In the case of the b.c.c. twins it is found that the migration energy may be up to 40% less than through the perfect crystal.[53] For twins and stacking faults in f.c.c. metals the effect is less marked but may be as high as 10%.[54] The reasons for this lowering of the migration energy depend on the details of the structure. In the case of b.c.c. twins the structure is more open so that the migrating atom, which hops from the new to the old vacancy site, has an easier route. In the f.c.c. case the reduction arises from the migrating atom being able to follow a convoluted path.

Similar calculations have been carried out for vacancy migration along dislocations in b.c.c. crystals.[48] Two edge dislocations with Burgers vector ½ [111] lying on (1$\bar{1}$0) and (11$\bar{2}$) slip planes have been considered for both iron and molybdenum. The maximum binding energy arises when the vacancy lies on the compression side of the core of the dislocation. A higher than average concentration of vacancies is therefore expected to occur at the dislocations. However in order to migrate along the dislocations the vacancies, or rather the associated migrating atoms, must pass through gaps between neighbouring atoms which are compressed. This process will therefore be more difficult than migration in the perfect crystal. Although the detailed results are different for the two dislocations and the two metals, due largely to the structures of the dislocations being very different, this general conclusion holds in all four cases. For the ½ [111] (1$\bar{1}$0) edge dislocation in iron the mechanism involved is illustrated by Fig. 13. The binding energies for the sites labelled A, B and C are 0.67 eV, 0.69 eV and 0.75 eV respectively and from symmetry D = B and C = E. As there is no close packed direction parallel to the dislocation line pipe diffusion of vacancies occurs along the path ABCDA' or the equivalent path ABEDA' marked by broken lines in Fig. 12(b). The migration energies for the jumps AB, BC, CD and DA' are 1.09 eV, 1.09 eV, 0.10 eV and 1.07 eV respectively compared with 0.68 eV for a perfect crystal. Therefore the additional energy of 0.41 eV required for this mechanism largely cancels the binding energy, so that pipe diffusion is little easier than bulk diffusion in this case. The effect is less marked for molybdenum and for the ½ [111] (11$\bar{2}$) dislocations. The mechanism also suggests that pipe diffusion of interstitials along the tension side of dislocations should be particularly favoured.

Relationships between theories of defects

The different theoretical approaches to obtaining an adequate understanding of the physical properties of defects in crystalline media all have their advantages and disadvantages. Thus the geometrical theories are very successful in characterising the general features of the various possible configurations of defects but can provide little information for example on the relative energies of these configurations. They do however form the basis upon which all other theories must develop. Continuum theories provide convenient methods for estimating the stress and displacement fields and hence the energies of defects. However they can give little information on the intricate structure of the core regions and hence the nucleation and close range interaction of defects. Thus in some applications involving long range effects continuum theories are adequate but in others the discrete nature of the material has to be taken into account. There are many ways in which this can be attempted involving different degrees of analytical sophistication, model complexity, computational dependence and reliability, and potential authenticity. The real-space simulation procedure relies heavily on computation based on realistic empirical inter-atomic potentials. It thus lacks the elegance and glamour of more analytical methods but provides a way of obtaining valuable results on many defects and on their interactions. However, developing the models and interpreting the results requires the use of both the geometric and the continuum approaches. Indeed interpretation of the mass of detailed information on atomic co-ordinates which is provided by this method is the most challenging and demanding aspect of the exercise. In some cases general rules are being established but unfortunately many defects appear to be highly individualistic with structures and properties which are difficult to predict from existing knowledge. It is in this area that it is hoped that related developments in non-local theories of material media can make a major contribution.

References

1. Henry, N.F.M. and Lonsdale, K., Eds., *International Tables for X-ray Crystallography, Vol. 1*, Kynoch Press, Birmingham, 1952.
2. Jaswon, M.A., *Introduction to Mathematical Crystallography*, Longmans, London, 1965.
3. Nye, J.F., *Physical Properties of Crystals*, Clarendon, Oxford, 1957.
4. Bilby, B.A. and Crocker, A.G., The theory of the crystallography of deformantion twinning, *Proc. Roy. Soc., Lond.*, A288, 240, 1965.
5. Hull, D., *Introduction to Dislocations*, Second edition, Pergamon, Oxford, 1975.
6. Weertman, J. and Weertman, J.R., *Elementary Dislocation Theory*, Macmillan, New York, 1964.
7. Hirth, J.P. and Lothe, J., *Theory of Dislocations*, McGraw-Hill, New York, 1968.
8. Nabarro, F.R.N., *Theory of Crystal Dislocations*, Clarendon, Oxford 1967.

9. Amelinckx, S., *The Direct Observation of Dislocations*, Academic Press, New York, 1964.

10. Crocker, A.G., Configurations of close-packed cluster of substitutional point defects in crystals, *Phil. Mag.*, 32, 379, 1975.

11. Crocker, A.G., Close packed clusters of five substitutional point defects in cubic crystals, *Crystal Lattice Defects*, 7, 239, 1978.

12. Crocker, A.G. and Faridi, B.A., Enumeration of migration, growth and contraction mechanisms for close-packed clusters of vacancies in fcc crystals, *J. Nuclear Materials*, 69-70, 671, 1978.

13. Malik, A.Q., Close packed clusters of point defects in nuclear materials, *M.Sc. thesis*, University of Surrey, Guildford, 1980.

14. Akhtar, J.I., Malik, A.Q. and Crocker, A.G., Enumeration of migration, growth and contraction mechanisms for clusters of vacancies and solutes in cubic crystals, in preparation.

15. Bilby, B.A., A rule for determining the displacements caused by the motion of a dislocation line, *Research*, 4, 387, 1951.

16. Bilby, B.A., Bullough, R. and Smith, E., Continuous distributions of dislocations: a new application of the methods of non-Riemannian geometry, *Proc. Roy. Soc. Lond.*, A231, 263, 1955.

17. Christian, J.W., *The Theory of Transformations in Metals and Alloys*, Second edition Part I, Pergamon, Oxford, 1975.

18. Christian, J.W. and Crocker, A.G., Dislocations and Lattice Transformations, in *Dislocations in Solids, Vol. 3, Moving Dislocations*, Nabarro, F.R.N., Ed., North Holland, Amsterdam, 1980, 165.

19. Crocker, A.G. and Flewitt, P.E.J., The migration of interphase boundaries by shear mechanisms, in *Interphase Boundaries in Solids*, Smith D.A. and Chadwick, G.A., Eds., Academic Press, London, in the press.

20. Bollmann, W., *Crystal Defects and Crystalline Interfaces*, Springer, Berlin, 1970.

21. Bevis, M. and Crocker, A.G., Twinning shears in lattices, *Proc. Roy. Soc. Lond.*, A304, 123, 1968.

22. Bevis, M. and Crocker, A.G., Twinning modes in lattices, *Proc. Roy. Soc. Lond.*, A313, 509, 1969.

23. Crocker, A.G., The crystallography of deformation twinning in alpha-uranium, *J. Nuclear Materials*, 16, 306, 1965.

24. Crocker, A.G. and Bevis, M., The crystallography of deformation twinning in titanium, in *The Science, Technology and Application of Titanium*, Jaffe R. and Promisel N. Eds., Pergamon, Oxford, 1970, 453.

25. Crocker, A.G., The crystallography of deformation twinning in alpha plutonium, *J. Nuclear Materials*, 41, 167, 1971.

26. Rechtien, J.J., Crocker, A.G. and Nelson, R.D., Twinning in alpha-neptunium, *J. Nuclear Materials*, 40, 134, 1971.

27. Acton, A.F., Bevis, M., Crocker, A.G. and Ross, N.D.H., Transformation strains in lattices, *Proc. Roy. Soc. Lond.*, A320, 101, 1970.

28. Crocker, A.G. and Singleton, G.A.A.M., The orientation dependence of the elastic moduli of crystalline mercury, *Phys. Stat. Solidi (a)*, 6, 635, 1971.

29. Singleton, G.A.A.M. and Crocker, A.G., The elastic energies of slip dislocations in crystalline mercury, *Phys. Stat. Solidi (a)*, 6, 645, 1971.

30. Tucker, M.O. and Crocker, A.G., The plane boundary in anisotropic elasticity, in *Mechanics of Generalized Continua*, Kröner, E., Ed., Springer, Berlin, 1968, 286.

31. Bacon, D.J., Barnett, D.M. and Scattergood, R.O., Anisotropic Continum Theory of Lattice Defects, *Prog. Mat. Sci.*, 23, 51, 1979.

32. Crocker, A.G. and Bacon, D.J., Elastic self-energies of undissociated dislocation jogs, *Phil. Mag.*, 15, 1155, 1967.

33. Heald, P.T., Discrete lattice models of point defects, in *Vacancies '76*, The Metals Society, London, 1977.

34. Kroner, E., *Kontinuumstheorie der Versetzungen und Eigenspannungen*, Springer, Berlin, 1958.

35. Bilby, B.A., Continuous distributions of dislocations, *Prog. Solid Mechanics*, 1, 331, 1960.

36. Datta Gairola, B.K. and Kröner, E., The nonlocal theory of elasticity and its application to interaction of point defects, in *Nonlocal Theories of Material Systems*, Polish Academy of Sciences, Warsaw, 1976, 5.

37. Rogula, D., Nonlocal models in elasticity, in *Nonlocal Theories of Material Systems*, Polish Academy of Sciences, Warsaw, 1976, 81.

38. Johnson, R.A., Interstitials and vacancies in α-iron, *Phys. Rev.*, 134A, 1329, 1964.

39. Crocker, A.G. and Bristowe, P.D., In-plane translations at crystalline interfaces, *Arch. Mech.*, 31, 3, 1979.

40. Crocker, A.G., Doneghan, M., and Ingle, K.W., The structure of small vacancy clusters in face-centred-cubic metals, *Phil. Mag. A*, 41, 21, 1980.

41. Bristowe, P.D. and Crocker, A.G., A computer simulation study of the structures of twin boundaries in body-centred cubic crystals, *Phil. Mag.*, 31, 503, 1975.

42. Fletcher, R. and Reeves, C.M., Function minimization by conjugate gradients, *Comp. J.*, 7, 149, 1964.

43. Ingle, K.W. and Crocker, A.G., A computer simulation study of the effect of applied stress on divacancy binding energies in body-centred cubic crystals, *Phys. Stat. Solidi (a)*, 38, 523, 1976.

44. Ingle, K.W. and Crocker, A.G., A computer simulation study of the migration of vacancies and divacancies in stressed body centred cubic metals, *J. Nuclear Materials*, 69-70, 667, 1978.

45. Crocker, A.G., Computer simulation of vacancy clusters in face-centred-cubic metals, in Interatomic Potentials and Crystalline Defects, Lee, J.K., Ed. TMS-AIME, Warrendale, 1981.

46. Bristowe, P.D. and Crocker, A.G., A computer simulation study of the structure of twinning dislocations in body-centred cubic metals, *Acta Metall.*, 25, 1363, 1977.

47. Ingle, K.W. and Crocker, A.G., The interaction between vacancies and the $\frac{1}{2}$ <111> $\{1\bar{1}0\}$ edge dislocation in body centred cubic metals, *Acta Metall.*, 26, 1461, 1978.

48. Miller, K.M., Ingle, K.W. and Crocker, A.G., A computer simulation study of pipe diffusion in body centred cubic metals, *Acta Metall.*, 29, 1599, 1981.

49. Crocker, A.G. and Faridi, B.A., Plane coalescence at grain boundaries, *Acta Metall.*, 28, 549, 1980.

50. Bristowe, P.D. and Crocker, A.G., The structure of high-angle (001) CSL twist boundaries in f.c.c. metals, *Phil. Mag. A*, 38, 487, 1978.

51. Ingle, K.W. and Crocker, A.G., On the structure of high-angle (110) CSL twist boundaries in f.c.c. metals, *Phil. Mag. A*, 41, 713, 1980.

52. Ingle, K.W., Bristowe, P.D. and Crocker, A.G., A computer simulation study of the interaction of vacancies with twin boundaries in body-centred cubic metals, *Phil. Mag.* 33, 663, 1976.

53. Ingle, K.W. and Crocker, A.G., Migration of vacancies near twin boundaries in body-centred-cubic metals, *Phil. Mag. A*, 37, 297, 1978.

54. Faridi, B.A. and Crocker, A.G., Migration of vacancies near stacking faults in face-centred-cubic metals, *Phil. Mag. A*, 41, 137, 1980.

THE NONLOCAL CONTINUUM THEORY OF LATTICE DEFECTS

B.K.D. Gairola

Institut für Theoretische und Angewandte Physik,

Pfaffenwaldring 57, 7000 Stuttgart 80, W-Germany

I. INTRODUCTION

Crystal imperfections, like those of human beings, come in seemingly endless variety. However, not all are equally interesting or easy to deal with. In these lectures we restrict ourselves to the most important kinds such as point defects and dislocations. Their presence has a profound effect on the physical properties of the material. Examples of point defects are vacancies and interstitials which are point defects consisting of the absence of an atom or the presence of an extra atom (Fig. 1). Dislocations, on the other hand, are line defects consisting of, for instance, an extra plane of atoms ending inside the crystal (edge dislocation, Fig. 2).

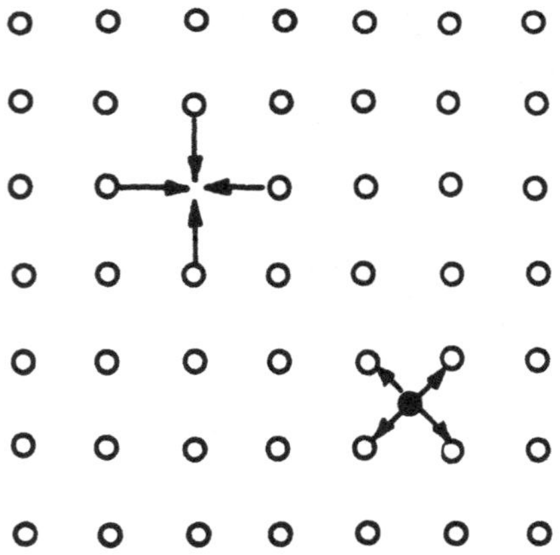

Figure 1

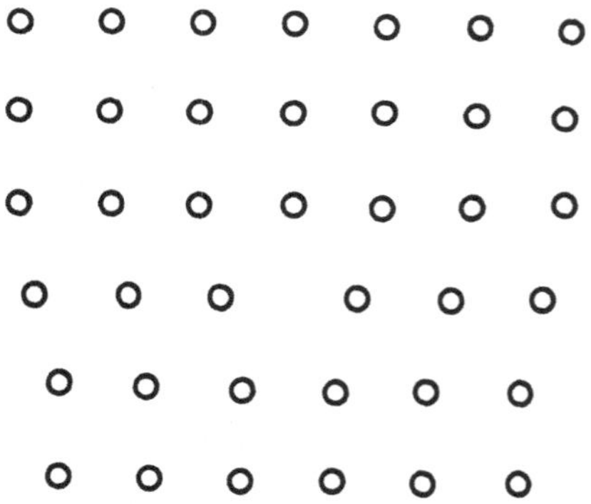

Figure 2

When these defects are introduced into a crystal lattice the host atoms are displaced from their original positions in the lattice. In principle, these displacements can be obtained from the lattice theory. In practice, however, these calculations would be too laborious and time-consuming and so it is of advantage to consider the crystal in its continuum limit.

The local continuum theory of lattice defects is well established and has been extensively used for calculations. Nevertheless, it has serious limitations. The domains in the local theory are so large as to make the macroscopic aspects of the crystalline body virtually insensitive to any microscopic details. Naturally, its application is limited to gross phenomena and a number of known effects elude it altogether.

The nonlocal theory which has been proposed independently by various authors incorporates the essential features of the lattice theory.[1-5] Thus it is more suitable for treating the lattice defects. It is the purpose of these lectures to illustrate that the nonlocal theory yields results strikingly different from those of the local theory and in better agreement with the lattice theory.

II. BASIC THEORY

1. Nonlocal theory of elasticity

The nonlocal theory of elasticity is an extension of the classical theory of elasticity intended to accomodate the essential features of a crystalline solid. Now what are the essential features of a crystal? It is in fact an assembly of discrete particles held together by forces of a certain range and arranged in a regular lattice structure.

Therefore, in any problem dealing with such bodies two characteristic parameters play a role: the range R of interaction and the discreteness length a. Obviously R can never be less than a $(R \leqslant a)$. Since a is a finite length we see that not only the long range of interaction but also the discreteness itself give rise to nonlocality.

However, in any experimental investigation of the properties of a crystal we use some apparatus which has a finite resolution. Thus there is always a lower limit to the length we can observe through the apparatus. Therefore the observable length λ is the third characteristic length which finally determines the degree of nonlocality in the theory.

Now let us consider these characteristic lengths one by one.

(a) Range of interaction

A long range of interaction implies that the disturbance in the neighbourhood of any point has an effect on all other points within the range of interaction. In particular we expect that the strain energy W is a generating functional depending on the strains $\varepsilon_{ij}(\underline{r})$ at all points. For small strains it can be expanded in a functional series. If the initial state is stress free and strains are so small that the terms of order hingher than second can be neglected we can write

$$W = \frac{1}{2} \int \int \varepsilon_{ij}(\underline{r}) \; C_{ijkl}(\underline{r},\underline{r}') \; \varepsilon_{kl}(\underline{r}') \, dv \, dv'. \quad (1.1)$$

The stress $\sigma_{ij}(\underline{r})$ is then defined as a variational derivative

$$\sigma_{ij}(\underline{r}) = \frac{\delta W}{\delta \sigma_{ij}(\underline{r})} = \int C_{ijkl}(\underline{r},\underline{r}') \; \varepsilon_{kl}(\underline{r}) \, dv'. \quad (1.2)$$

For a homogeneous medium

$$C_{ijkl}(\underline{r},\underline{r}') = C_{ijkl}(\underline{r}-\underline{r}') \quad (1.3)$$

and in the special case of a local continuum theory it has the form

$$C_{ijkl}\,(\underline{r}-\underline{r}') = C_{ijkl}\,\delta\,(\underline{r}-\underline{r}')\,. \tag{1.4}$$

In order to avoid writing all these integrals and indices we now introduce a compact notation which is very convenient in formal manipulations. We write the strain energy (1.1) in the form

$$W = \frac{1}{2}\,(\,\varepsilon\,|\,C\,|\,\varepsilon\,) \tag{1.5}$$

and the material law (1.2) as

$$|\,\sigma\,) = C\,|\,\varepsilon\,)\,. \tag{1.6}$$

Here $|\,\varepsilon\,)$ etc. are elements of a Hilbert space and C is an integral operator. Its kernel is the fourth rank tensor $C_{ijkl}(r,r')$. We shall adopt the following convention. Tensors of second rank will be denoted by lower case greek letters and vectors by lower case latin letters. Integral operators will be represented by upper case latin letters. The element $(\,|$ is called the dual of $|\,)$ and the product of $(\,|$ with $|\,)$ is identified with the scalar product $(\,|\,)$ which is the square of the norm of $|\,)$.

Given an operator A that acts on elements $|\,)$ one can define the action of the same operator on $(\,|$ by requiring that for any $|\alpha)$ and $(\beta|$ one has

$$\left[(\beta\,|\,A\right]\;|\alpha) \overset{\text{def}}{=} (\beta|\;\left[A\;|\alpha)\right] = (\beta\,|\,A\,|\,\alpha)\,. \qquad (1.7)$$

It should, however, be noted that $(\,|A$ is, in general, not the dual of $A\,|\,)$. Such a dual is given by $(\,|A^{+}$ where A^{+} is called the adjoint of operator A. It is defined by the relation

$$(\beta\,|\,A\,|\,\alpha) = (\alpha\,|\,A^{+}\,|\,\beta)\;. \qquad (1.8)$$

In this sense the operator C, as one can see from (1.1), is a self-adjoint operator, i.e. $C = C^{+}$. In fact all the operators which we will come across in this work will be self adjoint operators.

For a physically valid theory the operator C must satisfy certain other requirements. First of all it should be bounded. Secondly its inverse C^{-1} should also be bounded so that the relationship

$$|\,\varepsilon\,) = C^{-1}\,|\,\sigma\,)\,. \qquad (1.9)$$

which is inverse to (1.6) makes a physical sense.

It is tempting to consider C as a compact operator and in fact we ourselves have used a compact operator in our first paper.[3] The reason is that compact operators seem to be very obliging since they transform sequences that are merely bounded into "nicer" sequneces having convergent subsequences. This is a two-edged sword, however, since unavoidably there will be trouble with the inverse. The reason is that the inverse of a compact operator in an infinite dimensional space is unbounded. An operator C which strikes a better balance between itself and its inverse should have the kernel of the form

$$C_{ijkl}\,\delta(\underline{r}-\underline{r}') + C_{ijkl}(\underline{r}-\underline{r}') \qquad (1.10)$$

where the second kernel may define a compact operator.

The symmetry properties of an ideal crystal are characterized by certain groups of transformations which carry the crystal into a configuration which can not be distinguished from the original configuration. Therefore, it follows from the invariance of the strain energy that this $C_{ijkl}(r-r')$ should remain unchanged by such symmetry transformations. These symmetry transformations for various crystal classes can be described by orthogonal transformations. Suppose we apply the orthogonal transformation O which transforms the coordinate system x_i to

$$x_{i}' = \Omega_{ij} \, x_{j}' \tag{1.11}$$

and at the same time carries the point x_i to

$$\bar{x}_{i} = \Omega^{-1}_{ij} \, x_{j} \; . \tag{1.12}$$

Then putting $\underline{r}-\underline{r}' = \underline{R}$ we get

$$C_{i'j'k'l'}(\underline{\bar{R}}) = \Omega_{im}\,\Omega_{jn}\,\Omega_{kp}\,\Omega_{lq}\, C_{mnpq}\,(\Omega^{-1}_{rs}\, R_s)$$

$$= C_{ijkl}\,(\underline{\bar{R}}) \; . \tag{1.13}$$

Or in short notation

$$\sigma\, C\,(\sigma^{-1}\,\underline{R}) = C\,(\underline{R}) \tag{1.14}$$

(b) Discreteness

A discrete medium is most simply characterized by a cut-off length. That means we have to introduce some sort of truncating procedure. This can be done by using the quasi-continuum approach [1,4] or equivalently by using the sampling function approach [2] which is based on the sampling theorem introduced by Shannon [6] in information theory. This theorem says that if the Fourier transform $f(k)$ of a function $f(x)$ is equal to zero

$$f(k) = \text{F.T.}\,(f(x)) = 0 \quad \text{for } |k| > k_o \tag{1.15}$$

above a certain value of k then this function of x is uniquely determined from its discrete values

$$f_n = f_n(x) = f(na) \qquad (1.16)$$

where $a = \frac{\pi}{k_o}$.

In fact $f(x)$ is given by

$$f(x) = a \sum_{n=-\infty}^{\infty} f_n \, S(x-x_n) , \qquad -\infty < x < \infty \qquad (1.17)$$

where

$$S(x-x_n) = \frac{1}{2} \int_{-k_o}^{k_o} e^{ik(x-x_n)} \, dk = \frac{\sin k_o(x-x_n)}{k_o(x-x_n)} \qquad (1.18)$$

is called the sampling function. It is easy to show that $f(x) = f_n$ at $x = x_n$. In information theory this is sometimes called band-limited interpolation. In the same way the three dimensional sampling function in crystal physics is defined as

$$\delta_B(\underline{r}-\underline{r}_n) = \frac{1}{(2\pi)^3} \int_B \exp i\underline{k} \cdot (\underline{r}-\underline{r}_n) \; dv_k \qquad (1.19)$$

where $\underline{r}_n$ are the discrete lattice points and the integral is taken over the frist Brillouin zone. We now introduce a function of one continuous variable by the relation

$$f(\underline{r}) = v_o \sum_n f(\underline{r}_n) \, \delta_B(\underline{r}-\underline{r}_n) \qquad (1.20)$$

where v_o is the volume of an elementary cell. Similarly for a function of two variables we have

$$f(\underline{r},\underline{r}') = v_o^2 \sum_m \sum_n f(\underline{r}_m,\underline{r}_n)\, \delta_B(\underline{r}-\underline{r}_m)\, \delta_B(\underline{r}-\underline{r}_n)\,. \quad (1.21)$$

The inverse relations are given by

$$f(\underline{r}_n) = \int f(\underline{r})\, \delta_B(\underline{r}-\underline{r}_n)\, dv\,, \quad (1.22)$$

$$f(\underline{r}_m,\underline{r}_n) = \int f(\underline{r},\underline{r}')\, \delta_B(\underline{r}-\underline{r}_m)\, \delta_B(\underline{r}-\underline{r}_n)\, dv\, dv'\,. \quad (1.23)$$

The continuous function defined above takes precisely the values prescribed at the lattice points.

In this way the class of continuous functions representing the field variables is restricted to those functions which have their Fourier spectrum truncated at a certain value of the wave vector. In the language of distributions theory, these are tempered distributions having a compact support (Brillouin zone) in Fourier space. (For mathematical details see [7].) Such a distribution can also be considered as a convolution of a distribution with an arbitrary large support in the Fourier space and a distribution $\delta_B(\underline{r}-\underline{r}')$ defined by

$$\delta_B(\underline{r}-\underline{r}') = \frac{1}{(2\pi)^3} \int e^{i\underline{k}\cdot(\underline{r}-\underline{r}')} \, dv_k \; . \qquad (1.24)$$

This function plays the role of ordinary Dirac delta function in the quasi-continuum.

The quasi-local counterpart of (1.4) can now be written as

$$C(\underline{r}-\underline{r}') = C\,\delta_B(\underline{r}-\underline{r}') \; . \qquad (1.25)$$

It may be observed that this form satisfies the symmetry condition (1.14) whereas the ordinary delta function in (1.4) is invariant with respect to continuous group of rotations.

$\delta_B(\underline{r}-\underline{r}')$ for hexagonal, body centered cubic and face centered cubic lattices have been derived recently by Kotowski.[8] These have a rather complicated form. A simple form is obtained for simple cubic lattices given by

$$\delta_B(r-r') = \frac{1}{\pi^3} \prod_{i=1}^{3} \frac{1}{x_i-x_i'} \sin \frac{\pi(x_i-x_i')}{a} \qquad (1.26)$$

where a is the lattice constant (discreteness length).

Another simple case is that of Debye continuum which has been investigated by Kunin and coworkers.[9, 10] It corresponds to the case of a spherical Brillouin zone with radius

$k_B = \pi/a$. For this case the integration in (1.24) is easy to perform. We obtain

$$\delta_B(\underline{r}-\underline{r}') = \frac{1}{2\pi^2}\left[\frac{\sin k_B|\underline{r}-\underline{r}'|}{|\underline{r}-\underline{r}'|^3} - k_B\frac{\cos k_B|\underline{r}-\underline{r}'|}{|\underline{r}-\underline{r}'|^2}\right] . \quad (1.27)$$

The functions (1.26) and (1.27) go over to δ-functions for $k_B R \gg 1$.

For the two dimensional case δ_B has the form

$$\delta_B(\underline{r}-\underline{r}') = \frac{k_B}{2\pi|\underline{r}-\underline{r}'|} J_1(k_B|\underline{r}-\underline{r}'|) \quad (1.28)$$

where J_1 is the first order Bessel function.

In the space of the restricted class of functions introduced above one can define the Green functions of the Laplacian and the biharmonic operators in analogy to the ordinary Green functions.[9,10] Denoting the Green functions of the Laplacian and the biharmonic operator respectively by G_B and g_B we have

$$\Delta G_B(\underline{r}-\underline{r}') = \delta_B(\underline{r}-\underline{r}') , \quad (1.29)$$

$$\Delta\Delta g_B(\underline{r}-\underline{r}') = \delta_B(\underline{r}-\underline{r}') \quad (1.30)$$

where

$$G_B(\underline{R}) = \frac{1}{2\pi^2 R} S_i\, k_B R \quad (1.31)$$

and

$$g_B(\underline{R}) = \frac{1}{4\pi^2 k_B}\left[k_B\ R\ S_i\ k_B\ R + \frac{\sin k_B R}{k_B R} + \cos k_B\ R\right]. \tag{1.32}$$

Here we have again put $\underline{r}-\underline{r}' = \underline{R}$ and $R = |\underline{r}-\underline{r}'|$. S_i is the integral sine function.

For the two dimensional case which is of interest in certain dislocation problems we have

$$G_B(\underline{R}) = \frac{1}{2\pi} \quad \ln k_B\ L + \int_0^{k_B R} \frac{1}{t}\left[J_o(t) - 1\right] dt \tag{1.33}$$

and

$$g_B(\underline{R}) = \frac{1}{4\pi k_B^2}\left[J_o(k_B R) - \frac{1}{2} k_B R \cdot J_1(k_B R) + \frac{1}{2} k_B^2 R^2 + K_B\ R^2\ G_B(\underline{R})\right]. \tag{1.34}$$

Here J_o is the zeroeth order Bessel function and the constant L is the characteristic size of a sample material which is, generally, very large compared to the discreteness length.

(c) Observable length

The observable length can always be included in the theory by using it as the measuring unit. If this length is assumed to be very large in comparison to other lengths, the effective support of $C(\underline{r}-\underline{r}')$ can be looked upon as shrunk to a point. Then in the sense of distribution theory $C(\underline{r}-\underline{r}')$ can be expanded as a series in terms of delta function and its derivatives

$$C_{ijkl}(\underline{r}-\underline{r}') = \sum_{\alpha=0}^{\nu} C^{(\alpha)}_{ijklm_1\ldots m} \quad \partial_{m_1}\ldots\partial_{m_\alpha} \quad (\underline{r}-\underline{r}') \,. \tag{1.35}$$

This is the socalled weakly nonlocal case with the nonlocality of degree ν. It is equivalent to the usual long wave approximation.

Another way which is more exact is to consider the symmetry group which we can associate with the geometry of Brillouin zone or equivalently which satisfy the relation (1.14). It is then possible to expand such functions in terms of basis functions which transform according to the irreducible representation of the symmetry group. For instance, the functions forming the basis for an irreducible representation of the cubic symmetry group are frequently called kubic harmonics. They are linear combinations of spherical harmonics and were first introduced by von der Lage and Bethe.[11]

In order to keep the problems tractable we shall use in the following either the Debye continuum or the weakly nonlocal approximation. Moreover, we shall use the infinite medium approach neglecting the surface effects. These simplifications are not as drastic as they seem to be. They yield results which are qualitatively, and in some cases even quantitatively, nearer the results of the lattice theory.

2. The relationship between the force constants and the elastic constants

The relationship between the force constants of the lattice theory and $C(\underline{r}-\underline{r}')$ is rather complicated.[12] However, it is simpler to derive this relationship for the elastic constants $C^{(\alpha)}$ occurring in (1.35).

The lattice theory takes into account the long range of interaction from the very beginning. In this theory one assumes that the potential energy depends on all the atomic positions. For small displacements $\underline{u}(\underline{r}_n)$ of the atoms from their equilibrium position $\underline{r}_n$ one can expand the potential energy W in a series in powers of atomic displacements. If the body is in equilibrium and the displacements are so small that all the terms higher than second order can be neglected we can write

$$W = \frac{1}{2} \sum_m \sum_n \underline{u}(\underline{r}_m) \cdot \underline{\underline{\Phi}}(\underline{r}_m, \underline{r}_n) \cdot \underline{u}(\underline{r}_n) \tag{2.1}$$

where $\Phi(\underline{r}_m, \underline{r}_n)$ are the force constant tensors of second rank. This is usually called the harmonic approximation because it is equivalent to a set of independent harmonic oscillators. Using the relations (1.20) to (1.23) we can write (2.1) in the integral form

$$W = \frac{1}{2v_o^2} \ (u \mid \Phi \mid u) \ . \tag{2.2}$$

The kernel of the operator Φ depends only on the difference $\underline{r}-\underline{r}'$ for an infinite medium. Since the potential energy W is rotationally invariant whereas the displacements $\underline{u}(\underline{r})$ are not, the force constants $\underline{\underline{\Phi}}(\underline{r}-\underline{r}')$ must satisfy the condition

$$\int \Phi(\underline{r}-\underline{r}') \times \underline{r}' dv' = 0 \ . \tag{2.3}$$

If the body is displaced as a whole, i.e. given a constant displacement the potential energy remains the same. It follows that the force constants also satisfy the translational invariance condition

$$\int \Phi(\underline{r}-\underline{r}') \, dv' = 0 \ . \tag{2.4}$$

In order to compare the expression (2.2) with the elastic energy expression (1.5) we must further assume that the strains are determined by the displacements alone, i.e.

$$|\varepsilon) = \text{Def}\,|u) \ , \quad \text{or} \quad \varepsilon_{ij}(\underline{r}) = \frac{1}{2}(\partial_j u_i(\underline{r}) + \partial_i u_j(\underline{r})) \tag{2.5}$$

where the kernel of the operator Def is

$$\mathrm{Def}_{ijk}\,(\underline{r}-\underline{r}') = \frac{1}{2}\,(\,\delta_{jk}\,\partial_i + \delta_{ik}\,\partial_j)\,\delta(\underline{r}-\underline{r}')\,. \qquad (2.6)$$

Substituting (2.5) in (1.5) and using partial integration we obtain an expression of the form

$$W = \frac{1}{2}\,(\,u\,|\,L\,|\,u\,) \qquad (2.7)$$

where the kernel of the operator L is

$$\begin{aligned} L_{ik}(\underline{r}-\underline{r}') &= \partial_j\,\partial_l{}'\,C_{ijkl}\,(\underline{r}-\underline{r}') \\ &= -\,\partial_j\,\partial_l\,C_{ijkl}\,(\underline{r}-\underline{r}')\;. \end{aligned} \qquad (2.8)$$

It may be noticed that only that part of $C_{ijkl}(\underline{r}-\underline{r}')$ enters the expression on the right hand side of equation (2.8) which is symmetric in j and l. Let us call this part $T_{ikjl}(\underline{r}-\underline{r}')$, i.e.

$$T_{ikjl}\,(\underline{r}-\underline{r}') = \frac{1}{2}\left[C_{ijkl}\,(\underline{r}-\underline{r}') + C_{ilkj}\,(\underline{r}-\underline{r}')\right]. \qquad (2.9)$$

It may be easily verified that

$$T_{ikjl}\,(\underline{r}-\underline{r}') = T_{jlik}\,(\underline{r}-\underline{r}') \qquad (2.10)$$

and that $T_{ikjl}(\underline{r}-\underline{r}')$ is also symmetric in the indices i and k as well as j and l. The operator C can be expressed in terms of operator T in the following form

$$C = S\ T \tag{2.11}$$

where S is a tensor operator of rank eight. Its kernel is given by

$$S_{ijklmnpq}(\underline{r}-\underline{r}') = \Big[\delta_{im}\,\delta_{kn}\,\delta_{jp}\,\delta_{lq} + \delta_{jm}\,\delta_{kn}\,\delta_{ip}\,\delta_{lq} - \delta_{km}\,\delta_{ln}\,\delta_{ip}\,\delta_{jq}\Big]\,\delta(\underline{r}-\underline{r}')\,. \tag{2.12}$$

It is now clear that we must have

$$\partial_j\,\partial_l\,T_{ikjl}\,(\underline{r}-\underline{r}') = -\frac{1}{v_o^{\,2}}\,\Phi_{ik}\,(\underline{r}-\underline{r}')\,. \tag{2.13}$$

Expanding both sides of this equation in the same way as we did in (1.35) we obtain

$$\partial_j\,\partial_l \sum_{\alpha=o} T^{(\alpha)}_{ikjlm_1\ldots m_\alpha}\,\partial_{m_1\ldots\ m_\alpha}\,\delta(\underline{r}-\underline{r}')$$

$$= -\frac{1}{v_o^{\,2}} \sum_{\alpha=o} \Phi_{ikm_1\ldots m_\alpha}\,\partial_{m_1}\ldots\partial_{m_\alpha}\,\delta(\underline{r}-\underline{r}')\,. \tag{2.14}$$

Comparing the coefficients on both sides of (2.14) and keeping in mind the rotational and translational invarions conditions (2.3) and (2.4) we get the relations

$$T^{(\alpha)}_{ikjlm_1\ldots m_\alpha} = - \frac{1}{v_o^2} \Phi^{(\alpha+2)}_{ikjlm_1\ldots m_\alpha} \quad . \tag{2.15}$$

The coefficients on the right hand side are given by

$$\Phi^{(\alpha+2)}_{ikjlm_1\ldots m_\alpha} = \frac{(-1)^{\alpha+2}}{(\alpha+2)!} \int \Phi_{ik}(\underline{R})\, X_j X_l X_{m_1} \ldots X_m \, dV \tag{2.16}$$

where $\underline{R} = \underline{r}-\underline{r}'$ and $X_j = x_j - x_j'$ etc. In view of (1.26) and (2.11) we then have

$$C^{(\alpha)}_{ijklm_1\ldots m_\alpha} = - \frac{(-1)^{\alpha+2}}{(\alpha+2)!v_o^2} S_{ijklmnpq} \int \Phi_{mn}(R)$$

$$\cdot X_p X_q X_{m_1} \ldots X_{m_\alpha} \, dV \quad . \tag{2.17}$$

In centrosymmetric materials tensors of odd rank vanish. Let us consider the simplest case for such materials when

$$C_{ijkl}(\underline{r}-\underline{r}') = C^{(o)}_{ijkl}\, \delta(\underline{r}-\underline{r}') + C^{(2)}_{ijklmn}\, \partial_m \partial_n\, \delta(\underline{r}-\underline{r}') \, . \tag{2.18}$$

It follows from (2.17) that

$$C^{(o)}_{ijkl} = - \frac{1}{2v_o^2} S_{ijklmnpq} \int \Phi_{mn}(\underline{R})\, X_p X_q \, dV \quad , \tag{2.19}$$

$$C^{(2)}_{ijklmn} = - \frac{1}{24v_o^2} S_{ijklpqrs} \int \Phi_{pq}(\underline{R})\, X_r X_s X_m X_n \, dV \quad . \tag{2.20}$$

As an example let us calculate $C^{(o)}$ and $C^{(2)}$ for a face centered cubic lattice of special type considered by Hardy and Bullough.[13] In their calculation it was assumed that the interaction extends only to the nearest neighbours. In face centered cubic lattices there are twelve such neighbours. Therefore we must replace these integrals by summation over these twelve positions. Using the relations (1.22) and (1.23)

$$C^{(o)}_{ijkl} = - \frac{1}{2v_o} S_{ijklmnpq} \sum_{\underline{R}} \Phi_{mn}(\underline{R})\, X_p\, X_q \,, \tag{2.21}$$

$$C^{(2)}_{ijklmn} = - \frac{1}{24v_o} S_{ijklpqrs} \sum_{\underline{R}} \Phi_{pq}(\underline{R})\, X_r X_s X_m X_n \,. \tag{2.22}$$

In f.c.c. lattices it is easily shown by applying the symmetry operations that there are three independent force constants. If we take the origin at $\underline{r}$ and $\underline{r}'$ is the d(1,1,0) position where d is one half the cubic unit cell side we have

$$(000,110) = \begin{pmatrix} \alpha & \gamma & 0 \\ \gamma & \alpha & 0 \\ 0 & 0 & \beta \end{pmatrix} \,. \tag{2.23}$$

The force constant matrices for the remaining eleven first neighbours are obtained by simply using the appropriate transformations. For instance applying a counterclockwise rotation by 90° about [010] we get

$$(000,011) = \begin{pmatrix} \beta & 0 & 0 \\ 0 & \alpha & \gamma \\ 0 & \gamma & \alpha \end{pmatrix} . \tag{2.24}$$

Equivalently one can use the spectral form of

$$\Phi_{ik}(\underline{r}-\underline{r}') = \beta\,\delta_{ik} + \gamma h_i\, h_k + (\beta - \alpha + \gamma)(1-h_i^{\,2})(1-h_k^{\,2}) \tag{2.25}$$

where h is the unit vector in the direction of $\underline{R}$.

Hardy and Bullough not only assumed the nearest neighbour interaction but also that the restoring force acting against shear displacements, perpendicular to a nearest neighbour bond, is equal to that opposing the direct strechting of the bond. This assumption implies that $\alpha = \beta$ and $\gamma = 0$. In other words the force constant matrix is simply proportional to the unit tensor

$$\Phi_{ik} = \alpha\,\delta_{ik} . \tag{2.26}$$

Therefore, the expressions (2.21) and (2.22) for $C^{(o)}_{ijkl}$ and $C^{(2)}_{ijklmn}$ reduce to the following form

$$C^{(o)}_{ijkl} = -\frac{\alpha}{2v_o} S_{ijklmnpq} \sum_{\underline{R}} X_p\, X_q \quad , \tag{2.27}$$

$$C^{(2)}_{ijklmn} = -\frac{\alpha}{24v_o} S_{ijklrrpq} \sum_{\underline{R}} X_p X_q X_m X_n \quad . \tag{2.28}$$

In f.c.c. lattices

$$\sum_{\underline{R}} X_p X_q = \frac{a^2}{6} \delta_{pq}$$

where a is the length of the edge of the cubic unit cell that has four times the volume of the elementary cell. That means $v_o = a^3/4$. Hence

$$C^{(o)}_{ijkl} = - \frac{\alpha}{3a} S_{ijklmmnn}$$
$$= - \frac{\alpha}{3a} (\delta_{ik} \delta_{jl} + \delta_{jk} \delta_{il} - \delta_{ij} \delta_{kl}) . \quad (2.29)$$

Thus in Voigt notation we have

$$C^{(o)}_{11} = C^{(o)'}_{44} = - C^{(o)}_{12} = - \frac{\alpha}{3a} \quad (2.30)$$

and

$$C^{(o)}_{ijkl} = C_{11} (\delta_{ik} \delta_{jl} + \delta_{jk} \delta_{il} - \delta_{ij} \delta_{kl}) . \quad (2.31)$$

The fourth order nearest neighbour sum occurring in (2.28) has the symmetry of elastic constants obeying the Cauchy relations

$$\sum_{\underline{R}} X_p X_q X_m X_n = A(\delta_{pq} \delta_{mn} + \delta_{pm} \delta_{qn} + \delta_{pn} \delta_{qm}) + B \delta_{pqmn} \quad (2.32)$$

where

$$\delta_{pqmn} = \begin{cases} 1 & \text{for} \quad p = q = m = n \\ 0 & \text{otherwise} . \end{cases} \tag{2.33}$$

The constants A and B depend on the lattice types. They can be evaluated from the invariants

$$I_1 = X_1^2 X_2^2 + X_2^2 X_3^2 + X_3^2 X_1^2 = 3 A , \tag{2.34}$$

$$I_2 = X_1^4 + X_2^4 + X_3^4 = 9 A + 3 B . \tag{2.35}$$

For f.c.c. lattices

$$I_1 = \frac{a^4}{4} = \frac{1}{2} I_2 , \tag{2.36}$$

and so

$$A = \frac{a^4}{12} ; \quad B = - \frac{a^4}{12} . \tag{2.37}$$

Thus we obtain

$$\begin{aligned} C^{(2)}_{ijklmn} &= \frac{C^{(o)}_{11} a^2}{24} S_{ijklrrpq} (\delta_{pq} \delta_{mn} + \delta_{pm} \delta_{qn} + \delta_{pn} \delta_{qm} - \delta_{pqmn}) \\ &= \frac{C^{(o)}_{11} a^2}{24} \Big[(\delta_{ik} \delta_{jl} + \delta_{jk} \delta_{il} - \delta_{ij} \delta_{kl}) \delta_{mn} \\ &\quad + \delta_{ik} \delta_{jm} \delta_{ln} + \delta_{jk} \delta_{im} \delta_{ln} - \delta_{kl} \delta_{im} \delta_{jn} + \delta_{ik} \delta_{jn} \delta_{lm} + \delta_{jk} \delta_{in} \delta_{lm} \\ &\quad - \delta_{kl} \delta_{in} \delta_{jm} - \delta_{ik} \delta_{jlmn} - \delta_{jk} \delta_{ilmn} + \delta_{kl} \delta_{ijmn} \Big] . \end{aligned} \tag{2.38}$$

It can be seen that such a medium behaves isotropically in the long wave limit.

3. Equilibrium equation and its solution in terms of displacement Green function

The fundamental field equation for the time independent behaviour of an elastic body is the equilibrium equation which can be derived by using the principle of virtual work. This principle may be stated as follows: If an elastic body is in equilibrium under the action of certain body forces, and if a virtual displacement is given to each mass point, then the work done by the body forces acting through this displacement is equal to the increase in internal energy. In the present case it means

$$(\sigma \mid \text{Def} \mid \delta u) = (f \mid \delta u) \tag{3.1}$$

where $\delta \underset{-}{u}$ is the virtual displacement and $\underset{-}{f}$ is the body force density. Since $\delta \underset{-}{u}$ is arbitrary with certain restrictions this equation implies the equilibrium equation

$$\text{div} \mid \sigma) = \mid f) \, . \tag{3.2}$$

Alternatively we can also write (3.1) in the form

where the kernel of L is given by (2.8). The equilibrium equation thus takes the form

$$L\,|u) = |f)\,. \tag{3.3}$$

The solution of (3.3) can be formally written as

$$|u) = G\,|f)\,. \tag{3.4}$$

Here G is the Green function satisfying the relation

$$L\,G = 1 \tag{3.5}$$

where 1 is the identity operator. The product LG implies that its kernel is given by scalar multiplication as well as convolution, i.e. its kernel can be written as

$$\int L_{ik}\,(\underline{r}-\underline{r}')\,G_{kl}\,(\underline{r}'-\underline{r}'')\,dv'\,. \tag{3.6}$$

In an infinite medium we assume that $\underline{\underline{G}}$ vanishes at infinity. The kernel of the operator G which is a second rank tensor is formally given by

$$\underline{\underline{G}} = |\underline{\underline{L}}|^{-1}\,\gamma \tag{3.7}$$

where $\underline{\underline{L}}$ represents the kernel of L defined by (2.8) and where $|\underline{\underline{L}}|$ is the determinant defined by

$$|\underline{\underline{L}}| = \frac{1}{6}\,\underline{\underline{L}} \times\times \underline{\underline{L}} \cdot\cdot \underline{\underline{L}} \tag{3.8}$$

and γ is the cofactor written as

$$\underline{\underline{\gamma}} = \frac{1}{2} \underline{\underline{L}}^T \times \times \underline{\underline{L}}^T \qquad (3.9)$$

where $\underline{\underline{L}}^T$ is the transpose of $\underline{\underline{L}}$.

With a few exceptions, it is not possible to calculate G in this way. However, in most cases of interest, one can decompose L in two parts

$$L = L^o + \delta L \qquad (3.10)$$

in such a way that the Green function G^o corresponding to L^o can be calculated easily.

We now rewrite equation (3.5) in the form

$$(L^o + \delta L)\, G = 1 \qquad (3.11)$$

or

$$L^o\, G = 1 - \delta L\, G \qquad (3.12)$$

which yields

$$G = G^o - G^o\, \delta L\, G\,. \qquad (3.13)$$

This equation is generally known as Dyson's equation and may be put in the form

$$G = (1 + G^o\, \delta\, L)^{-1}\, G_o \qquad (3.14)$$

which is equivalent to the infinite series solution

$$G = G^o - G^o \delta L\, G^o + G^o \delta L\, G^o \delta L\, G^o - \dots \qquad (3.15)$$

which is known to mathematicians as Neumann series. This is just a formal solution and has a meaning only if the series converges or equivalently if the norm of $G^o \delta L$ is sufficiently small.

If the solution (3.14) is substituted on the right hand side of (3.13) we obtain

$$G = G^o - G^o\, T\, G^o \qquad (3.16)$$

where

$$T = \delta L\, (1 + G^o \delta L)^{-1} \; . \qquad (3.17)$$

We can also put

$$T = \delta L - \delta L\, G\, \delta L \; . \qquad (3.18)$$

This "T-matrix" is extensively used in the quantum scattering theory. Its importance lies in the fact that it completely separates the effect of the perturbed part of the operator L. This is evident if we use equations (3.3) and (3.4) to write the elastic energy in the form

$$(u\,|\,L\,|\,u) = (f\,|\,G\,|\,f) \; . \qquad (3.19)$$

Substituting (3.16) in the above we get

$$(f \mid G \mid f) = (u_o \mid L_o \mid u_o) - (u_o \mid T \mid u_o) \tag{3.20}$$

where

$$\mid u_o) = G^o \mid f) \quad . \tag{3.21}$$

The operator T itself can be written as a series

$$T = \delta L - \delta L\, G^o\, \delta L + \delta L\, G^o\, \delta L\, G^o\, \delta L - \ldots \tag{3.22}$$

As an example consider the weakly nonlocal case. If we take into account the first two terms in (1.35) we have

$$L^o_{ik}\ (\underline{r}-\underline{r}') = C^{(o)}_{ijkl}\, \partial_j\, \partial_l\, \delta(\underline{r}-\underline{r}') \quad , \tag{3.23}$$

$$\delta L_{ik}(\underline{r}-\underline{r}') = L^{(2)}_{ik}\ (\underline{r}-\underline{r}') = C^{(2)}_{ijklmn}\ \partial_j \partial_l \partial_m \partial_n\, \delta(\underline{r}-\underline{r}') \ . \tag{3.24}$$

Thus G is now given

$$G = G^o - G^o\ L^{(2)}\ G^o + G^o\ L^{(2)}\ G^o\ L^{(2)}\ G^o - \ldots \tag{3.25}$$

4. Modified Green functions and decomposition of general stress and strain fields

So far we have assumed that strains $\mid \varepsilon)$ can be defined uniquely in terms of $\mid u)$ in the form of $\mid \varepsilon) = \text{Def} \mid u)$. But it is well known that a continuous single-valued displacement field exists if, and only if, $\mid \varepsilon)$ satisfies the St. Venant's compatibility conditions which can be written as

$$\text{Inc } |\varepsilon) = 0 \,. \tag{4.1}$$

The kernel of incompatibility operator is given by

$$(\text{Inc})_{ijkl}(\underline{r}-\underline{r}') = \epsilon_{imk}\, \epsilon_{jnl}\, \partial_m\, \partial_n\, \delta(\underline{r}-\underline{r}') \tag{4.2}$$

where ϵ_{ijk} is the completely antisymmetric Levi Civita tensor. The operators Def and Inc were first introduced by Kröner.[14] He also pointed out the close analogy between these operators and the operators gradient and curl. For instance, the identity

$$\text{Inc Def} = 0 \tag{4.3}$$

can be considered as the counterpart of the identity

$$\text{curl grad} = 0 \,. \tag{4.4}$$

It is also easy to verify that

$$\text{div Inc} = 0 \tag{4.5}$$

and

$$\text{Inc grad} = 0 \,. \tag{4.6}$$

Equation (4.1) is not satisfied when the functions $|u)$ are not well defined. This is the case when strains are caused by local structural changes, for example, due to plastic deformation, recrystallization, phase transformations, thermal treatment etc. In general, therefore, a strain field can be decomposed into two parts

$$|\varepsilon) = |\overset{1}{\varepsilon}) + |\overset{2}{\varepsilon}) \tag{4.7}$$

where $|\overset{1}{\varepsilon})$ and $|\overset{2}{\varepsilon})$ satisfy the conditions

$$\mathrm{Inc}\,|\overset{1}{\varepsilon}) = 0 \;, \tag{4.8}$$

$$\mathrm{Inc}\,|\overset{2}{\varepsilon}) = |\eta) \;. \tag{4.9}$$

The function $|\eta)$ can be considered as the source function for $|\overset{2}{\varepsilon})$.

The general stress field $|\sigma)$ is accordingly decomposed into two parts

$$|\sigma) = |\overset{1}{\sigma}) + |\overset{2}{\sigma}) \tag{4.10}$$

where

$$|\overset{1}{\sigma}) = C\,|\overset{1}{\varepsilon}) \;, \tag{4.11}$$

$$|\overset{2}{\varepsilon}) = C\,|\overset{2}{\varepsilon}) \;. \tag{4.12}$$

We will now demonstrate that $|\overset{1}{\varepsilon})$ and $|\overset{2}{\varepsilon})$ are orthogonal with respect to the weight C, or equivalently $|\overset{1}{\sigma})$ and $|\overset{2}{\sigma})$ are orthogonal to $|\overset{2}{\varepsilon})$ and $|\overset{1}{\varepsilon})$ respectively. To this end we use the method of projection operators introduced by Kunin.[15] We first apply the operator Def to the equation (3.4) and thus obtain

$$|\overset{1}{\varepsilon}) = \mathrm{Def}\ G\,|f) \ . \qquad (4.13)$$

If we substitute $|f) = -\mathrm{div}\ |\sigma)$ in the above we get

$$|\overset{1}{\varepsilon}) = H\,|\sigma) \qquad (4.14)$$

where

$$H = \mathrm{Def}\ G\ \mathrm{Def}\ . \qquad (4.15)$$

Following Kröner we call H the modified strain Green function.[16] Its kernel is given by

$$H_{ijkl}(\underline{r}-\underline{r}') = \partial_j\,\partial_l\,G_{ik}(\underline{r}-\underline{r}')\Big|_{(ij),(kl)} \ . \qquad (4.16)$$

Here (ij) and (kl) imply symmetrization with respect to the enclosed indices. In view of (1.6) we can write (4.14) in the form

$$|\overset{1}{\varepsilon}) = \overset{1}{P}\,|\varepsilon) \qquad (4.17)$$

where

$$\overset{1}{P} = H\,C\ . \qquad (4.18)$$

Obviously $|\overset{2}{\varepsilon})$ can be put as

$$|\overset{2}{\varepsilon}) = |\varepsilon) - |\overset{1}{\varepsilon}) = \overset{2}{P}\,|\varepsilon) \qquad (4.19)$$

with

$$\overset{2}{P} = I - \overset{1}{P} = I - H\,C \qquad (4.20)$$

where I is the unit operator in the space of symmetric second rank tensor functions. Its kernel is given by

$$I_{ijkl}(\underline{r}-\underline{r}') = \frac{1}{2} (\delta_{ik} \delta_{jl} + \delta_{il} \delta_{jk}) \delta(\underline{r}-\underline{r}') , \quad (4.21)$$

By direct integration one can verify that

$$H\,C\,H = H \quad (4.22)$$

which was first proved by Dederichs and Zeller.[17] Using this relation we obtain

$$\overset{1}{P}\,\overset{1}{P} = H\,C\,H\,C = H\,C = \overset{1}{P} , \quad (4.23)$$

$$\overset{1}{P}\,\overset{2}{P} = H\,C\,(I - H\,C) = 0 , \quad (4.24)$$

$$\overset{1}{P} + \overset{2}{P} = I . \quad (4.25)$$

Hence $\overset{1}{P}$ and $\overset{2}{P}$ satisfy all the properties of orthogonal projection operators.

It is now straightforward to derive the following results for the stresses $|\overset{1}{\sigma})$ and $|\overset{2}{\sigma})$

$$|\overset{1}{\sigma}) = C\,H\,C\,|\varepsilon) \quad (4.26)$$

$$|\overset{2}{\sigma}) = (C - C\,H\,C)\,|\varepsilon) = F\,|\varepsilon) . \quad (4.27)$$

The relation (4.27) is the internal stress counterpart of (4.14). Hence it is natural to call F the modified stress Green function. Substituting (1.6) in (4.26) and

$$|\varepsilon) = S\,|\sigma)\,, \qquad (S = C^{-1}) \tag{4.28}$$

in (4.27) we obtain

$$|\overset{1}{\sigma}) = C\,H\,|\sigma)\,; \tag{4.29}$$

$$|\overset{2}{\sigma}) = F\,S\,|\sigma)\,. \tag{4.30}$$

It is easily verfied that the modified Green functions H and F are self-adjoint, i.e. their kernels satisfy the conditions

$$H_{ijkl}(\underline{r}-\underline{r}') = H_{klij}(\underline{r}-\underline{r}')\,, \tag{4.31}$$

$$F_{ijkl}(\underline{r}-\underline{r}') = F_{klij}(\underline{r}-\underline{r}')\,. \tag{4.32}$$

Since this property is also shared by C and S we find that

$$|\overset{1}{\sigma}) = \overset{1}{P}{}^{T}\,|\sigma)\,, \tag{4.33}$$

$$|\overset{2}{\sigma}) = \overset{2}{P}{}^{T}\,|\sigma)\,. \tag{4.34}$$

Here $\overset{1}{P}{}^{T}$ and $\overset{2}{P}{}^{T}$ are transposes of $\overset{1}{P}$ and $\overset{2}{P}$ respectively. Obvously we have also the relations

$$(\overset{1}{\sigma}| = (\sigma|\,\overset{1}{P}\,, \qquad (\overset{2}{\sigma}| = (\sigma|\,\overset{2}{P}\,; \tag{4.35}$$

$$(\overset{1}{\varepsilon}| = (\varepsilon|\,\overset{1}{P}{}^{T}\,, \qquad (\overset{2}{\varepsilon}| = (\varepsilon|\,\overset{2}{P}{}^{T}\,. \tag{4.36}$$

We now find that

$$(\overset{1}{\sigma} \mid \overset{2}{\varepsilon}) = (\sigma \mid \overset{1}{P} \overset{2}{P} \mid \varepsilon) = 0 , \qquad (4.37)$$

$$(\overset{2}{\sigma} \mid \overset{1}{\varepsilon}) = (\sigma \mid \overset{2}{P} \overset{1}{P} \mid \varepsilon) = 0 \qquad (4.38)$$

or equivalently

$$(\overset{1}{\varepsilon} \mid C \mid \overset{2}{\varepsilon}) = (\overset{2}{\varepsilon} \mid C \mid \overset{1}{\varepsilon}) = 0 \qquad (4.39)$$

which proves our assertions. If we rewrite (4.38) as

$$(\overset{2}{\sigma} \mid \overset{1}{\varepsilon}) = (\overset{2}{\sigma} \mid \mathrm{Def} \mid u) = 0$$

and apply partial integration we get

$$\mathrm{div} \mid \overset{2}{\sigma}) = 0 \qquad (4.40)$$

We can conclude that stress $\mid \overset{2}{\sigma})$ can exist without the action of external forces. For this reason stresses of this type are called internal stresses and they are associated with incompatible strains. In fact, the incompatibilities $\mid \eta)$ can be considered as sources of internal stresses.

The modified Green functions H and F can not be solved directly for the general case. We can obtain perturbation solutions in terms of simplest H^o and F^o exactly on the same lines as for the displacement Green function G.

For instance if we decompose C in two parts

$$C = C^{o} + \delta C \tag{4.41}$$

and keep in mind that both $C\,H$ and $C^{o}\,H^{o}$ define the same projection operator we can write

$$(C^{o} + \delta C)\,H = C^{o}\,H^{o}. \tag{4.42}$$

Multiplying both sides by H^{o} from the left and using (4.22) we obtain

$$H = H^{o} - H^{o}\,\delta C\,H. \tag{4.43}$$

The formal solution of this equation is

$$H = (I + H^{o}\,\delta C)^{-1}\,H^{o}. \tag{4.44}$$

We can also write this equation in terms of a T_1-matrix

$$H = H^{o} - H^{o}\,T_1\,H^{o} \tag{4.45}$$

where

$$T_1 = \delta C - \delta C\,H\,\delta C = C\,(I + H^{o}\,\delta C)^{-1}. \tag{4.46}$$

In the same way equations can be obtained for the modified Green function F as follows

$$F = (I + F^{o}\,\delta S)^{-1}\,F^{o} = F^{o} - F^{o}\,T_2\,F^{o} \tag{4.47}$$

where

$$T_2 = S\,(I + F^{o}\,\delta S)^{-1}. \tag{4.46}$$

5. Incompability and internal stress

In the local theory Kröner developed an elegant method to cope with stresses and energies introduced into the material by sources of internal stress.[18,19] His method was generalized to the nonlocal case by Kunin.[15] In this theory the governing field equations are

$$\mathrm{Inc}\,|\varepsilon) = |\eta)\ , \tag{5.1}$$

$$|\varepsilon) = S\,|\sigma)\ , \tag{5.2}$$

$$|\sigma) = F\,|\varepsilon)\ , \tag{5.3}$$

$$\mathrm{div}\,|\sigma) = 0\ . \tag{5.4}$$

If we apply the operator Inc to (5.2) and use (5.1) we obtain

$$\mathrm{Inc}\,S\,|\sigma) = |\eta)\ . \tag{5.5}$$

On the other hand, in view of (4.5) and (5.4) one can put

$$|\sigma) = \mathrm{Inc}\,|\psi) \tag{5.6}$$

where $|\psi)$ is a symmetric tensor of second rank. It is known as stress function. Hence (5.5) can be rewritten as

$$D\,|\psi) = |\eta) \tag{5.7}$$

where

$$D = {}'\mathrm{Inc}\,S\,\mathrm{Inc}\ . \tag{5.8}$$

Note that D is a fourth rank tensor operator.

We now introduce yet another Green function J defined by

$$D\,J = I \,. \tag{5.9}$$

The solution of (5.7) is then given by

$$|\psi) = J\,|\eta) \,. \tag{5.10}$$

In view of (5.1) and (5.6) we obtain

$$|\sigma) = \mathrm{Inc}\,J\,|\eta) = \mathrm{Inc}\,J\,\mathrm{Inc}\,|\varepsilon) \,. \tag{5.11}$$

Comparing it with (5.3) we have

$$F = \mathrm{Inc}\,J\,\mathrm{Inc} \,. \tag{5.12}$$

From (5.8) and (5.9) we obtain

$$\mathrm{Inc}\,S\,F = \mathrm{Inc}\,S\,\mathrm{Inc}\,J\,\mathrm{Inc} = D\,J\,\mathrm{Inc} = \mathrm{Inc} \,. \tag{5.13}$$

This is also obvious from the fact that $S\,F = \overset{2}{P}$ which is the unit operator in the subspace of incompatible strains. It is interesting to note that the elastic energy of an internal stress system can be expressed as

$$\frac{1}{2}(\sigma\,|\,\varepsilon) = \frac{1}{2}(\varepsilon|F|\varepsilon) = \frac{1}{2}(\eta|J|\eta) \,. \tag{5.14}$$

In general, it is difficult to obtain J. However, for the local isotropic case a clever procedure was introduced by Kröner[18] and independently by Marguerre[20] which makes it possible to calculate J in a simple way. We shall generalize it to the nonlocal isotropic case. For the sake of simplicity we consider the kernel of C in the form (1.25). Then for the isotropic medium it can be written as

$$C_{ijkl}(\underline{r}-\underline{r}') = \left[\lambda\delta_{ij}\delta_{kl} + \mu(\delta_{ik}\delta_{jl} + \delta_{il}\delta_{jk})\right]\delta_B(\underline{r}-\underline{r}'). \tag{5.15}$$

Hence the kernel of S is given by

$$S_{ijkl}(\underline{r}-\underline{r}') = \left[\frac{1}{2\mu}(\delta_{ik}\delta_{jl} + \delta_{il}\delta_{jk}) - \frac{\lambda}{2\mu(3\lambda+2\mu)}\delta_{ij}\delta_{kl}\right] \cdot \delta_B(r-r') . \tag{5.16}$$

We now introduce a modified function

$$|\chi) = B\,|\psi) \tag{5.17}$$

where the kernel of B is

$$B_{ijkl}(\underline{r}-\underline{r}') = \frac{1}{4\mu}\left(\delta_{ik}\delta_{jl} + \delta_{il}\delta_{jk} - \frac{\lambda}{\lambda+2\mu}\delta_{ij}\delta_{kl}\right) \cdot \delta_B(\underline{r}-\underline{r}') . \tag{5.18}$$

The inverse relation is given by

$$|\psi) = B^{-1}\,|\chi) \tag{5.19}$$

with

$$B^{-1}_{ijkl}(\underline{r}-\underline{r}') = \mu\,(\delta_{ik}\,\delta_{jl} + \delta_{il}\,\delta_{jk} + \frac{2\lambda}{\lambda+2\mu}\,\delta_{ij}\,\delta_{kl})\,\delta_B(\underline{r}-\underline{r}'). \tag{5.20}$$

Thus equation (5.7) takes the form

$$D'\,|\,\chi) = |\,\eta) \tag{5.21}$$

where

$$D' = D\,B^{-1}\;. \tag{5.22}$$

The kernel of D' has the form

$$D'_{ijkl}(\underline{r}-\underline{r}') = \Big[\Delta\Delta\, I_{ijkl}(\underline{r}-\underline{r}') - (\mathrm{Def})_{ijk}\,\partial_l\,\Delta + \frac{2(\lambda+\mu)}{3\lambda+2\mu} \cdot \partial_i\partial_j\partial_k\partial_l + \frac{\lambda}{3\lambda+2\mu}\,\delta_{ij}\partial_k\partial_l\Delta\Big]\,\delta_B(\underline{r}-\underline{r}'). \tag{5.23}$$

If we now take the gauge condition

$$\mathrm{div}\;|\,\chi) = 0 \tag{5.24}$$

equation (5.21) gets much simpler, i.e.

$$\Delta\Delta\,|\,\chi) = |\,\eta)\;. \tag{5.25}$$

The fact that we can always find a $|\,\chi)$ with zero divergence is shown as follows. Suppose we found a solution $|\,\chi')$ of (5.21) which does not satisfy (5.24). The function

$$|\,\chi) = |\,\chi') + \mathrm{Def}\;|\,\varphi) \tag{5.26}$$

then will give the same value of $|\sigma)$ in (5.6) by virtue of (4.3). Now if (5.24) must hold for $\chi)$ then

$$\operatorname{div} |\chi') + \operatorname{div} \operatorname{Def} |\varphi) = 0 \tag{5.27}$$

or

$$\operatorname{div} \operatorname{Def} |\varphi) = - \operatorname{div} |\chi') \ . \tag{5.28}$$

Since $|\chi')$ is known we can calculate $|\varphi)$. We can then use (5.26) to find $|\chi)$ which satisfies (5.24). Thus we obtain (5.25). Equation (5.7) now becomes

$$B\ |\psi) = |\eta) \tag{5.29}$$

for which the Green function J is relatively easy to obtain. For instance, consider the case when the medium is a Debye continuum for which the Brillouin delta function δ_B has the form (1.22). In this case kernel of J can be written as

$$J_{ijkl}(\underline{r}-\underline{r}') = \mu(\delta_{ik}\delta_{jl} + \delta_{il}\delta_{jk} + \frac{2\lambda}{\lambda+2\mu}\delta_{ij}\delta_{kl})g_B(\underline{r}-\underline{r}') \tag{5.30}$$

where $g_B(\underline{r}-\underline{r}')$ is given by the equation (1.32).

For more general cases we obtain J as usual by perturbation procedure, e.g.

$$J = (I + J^o\ S)^{-1}\ J^o = J^o - J^o\ T_3\ J^o \tag{5.31}$$

where J^o may be given by (5.30) and

$$T_3 = \delta S\,(I + J^o\,\delta S)^{-1} \quad . \tag{5.32}$$

6. Interaction energy between stress systems

Suppose that in the body there exist two stresses $|\sigma^A)$ and $|\sigma^B)$ and the corresponding strains $|\varepsilon^A)$ and $|\varepsilon^B)$. Since we are restricting attention to the linear theory of elasticity, the total stress and strain in the material is the sum of the contributions from each system

$$|\sigma) = |\sigma^A) + |\sigma^B) \quad , \tag{6.1}$$

$$|\varepsilon) = |\varepsilon^A) + |\varepsilon^B) \quad . \tag{6.2}$$

The interaction energy between the two systems is then defined as the difference between the elastic energy when both systems coexist and the sum of the individual energies when each alone exists, i.e.

$$W_{int} = \frac{1}{2}\ (\sigma^A + \sigma^B\ |\ \varepsilon^A + \varepsilon^B) - (\sigma^A\ |\ \varepsilon^A) - (\sigma^B\ |\ \varepsilon^B)$$

$$= \frac{1}{2}\ (\sigma^A\ |\ \varepsilon^B) + (\sigma^B\ |\ \varepsilon^A)\ . \tag{6.3}$$

If the elastic operator C is not changed then in view of its symmetry we can write

$$W_{int} = \frac{1}{2}\ (\varepsilon^A\ |\ C\ |\ \varepsilon^B) + (\varepsilon^B\ |\ C\ |\ \varepsilon^A)$$

$$= (\varepsilon^A\ |\ C\ |\ \varepsilon^B) = (\sigma^A\ |\ \varepsilon^B)\ . \tag{6.4}$$

Now let us consider the case that $|\sigma^A)$ and $|\sigma^B)$ satisfy the equilibrium equations

$$\mathrm{div}\,|\sigma^A) = |f^A) \quad , \tag{6.5}$$

$$\mathrm{div}\,|\sigma^B) = |f^B) \quad . \tag{6.6}$$

Accordingly we have

$$|\varepsilon^A) = \mathrm{Def}\,|u^A) \quad , \tag{6.7}$$

$$|\varepsilon^B) = \mathrm{Def}\,|u^B) \quad . \tag{6.8}$$

Using these relations and partial integration the interaction energy can be written as

$$W_{int} = - (f^A \lceil u^B) \tag{6.9}$$

or in terms of Green function G as

$$W_{int} = - (f^A | G | f^B) \quad . \tag{6.10}$$

Now suppose one of the stress systems, say $|\sigma^A)$, is an internal stress, i.e. satisfies the equation

$$\mathrm{div}\,|\sigma^A) = 0 \quad .$$

Using the relation (5.6) we can write

$$W_{int} = (\sigma^A \mid \varepsilon^B) = (\psi^A \mid Inc \mid \varepsilon^B) \quad . \qquad (6.11)$$

In view of the compatibility condition (4.1) we see at once that the interaction energy between a system of internal stress and a system of external stress vanishes. This is the well known Colonetti theorem.[21]

The total energy is now simply the sum of individual energies or self energies,

$$W = \frac{1}{2} (\sigma^A \mid \varepsilon^A) + \frac{1}{2} (\sigma^B \mid \varepsilon^B) \quad . \qquad (6.12)$$

Making use of (4.14) and (4.25) we have

$$W = \frac{1}{2} (\varepsilon^A \mid F \mid \varepsilon^A) + \frac{1}{2} (\sigma^B \mid H \mid \sigma^B)$$

$$= \frac{1}{2} (\eta^A \mid J \mid \eta^A) + \frac{1}{2} (\sigma^B \mid H \mid \sigma^B) \qquad (6.13)$$

which nicely demonstrates the role played by modified Green functions. The first terms in the equations above represent the self energy of an internal stress system.

If both systems are internal stresses their interaction is given by

$$W_{int} = (\varepsilon^A \mid F \mid \varepsilon^B) = (\eta^A \mid J \mid \eta^B) \quad . \qquad (6.14)$$

For the more general case we can write $C = C^o + \delta C$ and accordingly we obtain

$$W_{int} = (\varepsilon^A | F^o | \varepsilon^B) - (\varepsilon^A | F^o T_2 F^o | \varepsilon^B) \qquad (6.15)$$

and similar other forms corresponding to G, H and J.

7. Defects considered from two alternative viewpoints

Defects in a continuum can be considered from two different view points. According to the point of view customary in the lattice theory if one introduces a defect into an otherwise perfect crystal it has two effects:

(1) It will exert forces on the neighbouring atoms of the host crystal. This is the socalled size effect. It can be described in the continuum theory by a force distribution $f(\underline{r})$ which must satisfy the equilibrium equation

$$\int f(\underline{r}) \, dv = 0 \qquad (7.1)$$

$$\int f(\underline{r}) \times \underline{r} \, dv = 0 \quad . \qquad (7.2)$$

(2) In the close vincinity of the defect the force constants are modified due to the changed coupling between the atoms. This is known as inhomogeneity effect. In the continuum theory this effect can be described by a more general operator C'. Its kernel is given by

$$C'(r,r') = C(r-r') + \delta C(r,r') \tag{7.3}$$

where δC is non-zero only in a small region. Correspondingly we can define the operator L' which has the form

$$L' = L + \delta L \, , \tag{7.4}$$

With this point of view the displacement Green function method described in section 3 is a natural choice for treating defects.

Alternatively, as is more common in continuum theory we can associate with the defect a suitable state of internal stress in the following way. Suppose we cut out a small volume element from the continuum with elastic constant C and replace it by another piece of material with elastic constant $C + \delta C$ and subject it to stress-free incompatible strain $\underline{\underline{\varepsilon}}^p$. It will obviously not fit into the hole due to incompatibility. We then apply suitable forces to this volume element producing an elastic deformation $-\underline{\underline{\varepsilon}}^p$, which brings it back to the original shape to fit the hole perfectly. We now replace the strained volume element back in the hole and cement the material together and thus obtain a continuous medium with a ditribution of forces. If we now remove these forces the medium would undergo an elastic deformation $\underline{\underline{\varepsilon}}^e$ reaching a final state which is in elastic equilibrium with no body forces present. Thus the corresponding stress is internal. We also see that in the end the total strain is given by

$$\underline{\underline{\varepsilon}}^{t} = - \underline{\underline{\varepsilon}}^{p} + \underline{\underline{\varepsilon}}^{e} . \qquad (7.5)$$

Since the body in the natural state as well as at the end remains compact $\underline{\underline{\varepsilon}}^{t}$ is a compatible strain. Hence it follows that

$$- \mathrm{Inc} | \varepsilon^{e}) = \mathrm{Inc} | \varepsilon^{p}) = | \eta) . \qquad (7.6)$$

We thus find that the stress associated with a defect is internal stress and so the method discussed in section 5 becomes applicable.

In the following we shall illustrate both methods. The displacement Green function method will be applied to point defects and dislocations will be treated by the method of internal stresses.

8. Elastic Interaction between point defects

If more than one defect are present in the crystal they can interact in various ways. However, the interaction which takes place through the strain or displacement field produced by defects when the lattice is allowed to relax is always present. This type of interaction is called the elastic interaction. It plays the major role in many problems of great interst such as x-ray scattering, internal friction phenomena, aggregation of defects and various diffusion processes.

We consider here the case of two defects situated at two different sites, separated by a distance large enough so that the corresponding nonlocal force distributions do not overlap. The defect operator L of such a system is assumed to be

$$L' = L + \delta L^A + \delta L^B \tag{8.1}$$

where L is the operator corresponding to the homogeneous medium.

We define further

$$L^A = L + \delta L^A , \tag{8.2}$$

$$L^B = L + \delta L^B . \tag{8.3}$$

Hence

$$L' = L^A + \delta L^B = L^B + \delta L^A . \tag{8.4}$$

We denote G', G^A, G^B and G the Green functions corresponding to L', L^A, L^B and L. By considering L^A (or L^B) as the unperturbed part and δL^B (or δL^A) as the perturbation we obtain

$$G' = G^A - G^A T^B G^A = G^B - G^B T^A G^B \tag{8.5}$$

where

$$T^A = L^A (1 + G^B \delta L^A)^{-1} , \tag{8.6}$$

$$T^B = L^B (1 + G^A \delta L^B)^{-1} . \tag{8.7}$$

We have shown in section 6 that the interaction energy between two strain systems is given by $-(f^A | G | f^B)$ (see equation (6.10)). It was derived under the tacit assumption that the operator L is not perturbed by the two systems which is not valid in the present case. We again define the interaction energy between the two defects as the difference between the energy of both defects together and the sum of their individual energies, i.e.

$$W_{int} = \frac{1}{2} (f^A + f^B | G' | f^A + f^B) - (f^A | G^A | f^A) - (f^B | G^B | f^B) . \tag{8.8}$$

Since

$$G' - G^A = - G^A T^B G^A , \tag{8.9}$$

$$G' - G^B = - G^B T^A G^B , \tag{8.10}$$

we have

$$W_{int} = \frac{1}{2} (f^A | G^A T^B G^A | f^A) + (f^B | G^B T^A G^B | f^B) - (f^A | G | f^B) - (f^B | G' | f^A) . \tag{8.11}$$

Finally we can put the interaction energy in the form

$$W_{int} = W_1 + W_2 \quad .$$

Here W_1 is given by

$$W_1 = -\frac{1}{2}(f^A \mid G \mid f^B) - \frac{1}{2}(f^B \mid G \mid f^A) \quad . \tag{8.12}$$

It described the direct interaction of force distributions. The second term W_2 is the induced or inhomogeneity interaction given by

$$W_2 = \frac{1}{2} \quad (f^A \mid G^A \; T^B \; G^A \mid f^A) + (f^B \mid G^B \; T^A \; G^B \mid f^B) \; . \tag{8.13}$$

These are exact expressions but difficult to calculate in their full generality. However, in most problems of physical interest one can fairly assume that the defect interacts directly with its near neighbours so that δC has only a short range. In other words, the defect can be assumed to be localized in a small region. For practical purpose, therefore, we can replace the defect Green functions G', G^A, and G^B by the infinite medium Green function G. Thus we neglect extremely small terms of the order $(a/R)^3$ where a is the lattice constant and R is the typical crystal diameter. Furthermore if the perturbation δC is small

$$T^{A,B} \approx \delta L^{A,B} \quad .$$

In this way we get

$$W_1 = -\frac{1}{2} \quad (f^A \mid G \mid f^B) + (f^B \mid G \mid f^A) \quad , \tag{8.14}$$

$$W_2 = \frac{1}{2} \quad (f^A \mid G \,\delta L^B\, G \mid f^A) + (f^B \mid G \,\delta L^A\, G \mid f^B) \quad . \tag{8.15}$$

If the two defects are similar, that is they are characterized by similar force distribution and similar inhomogeneity, we can drop the labels and write

$$W_1 = -\,(f \mid G \mid f) \quad , \tag{8.16}$$

$$W_2 = (f \mid G \,\delta L\, G \mid f) \quad . \tag{8.17}$$

If the defects are separated by a large distance we can apply the weakly nonlocal approximation. In this case the kernel of δC can be considered as the delta function type. Furthermore, the force distribution can be expanded in terms of delta function and its derivation

$$f_1(\underline{r}) = \sum_{\alpha=o} P^{(\alpha)}_{ij_1\dots j_\alpha} \; \partial_{j_1\dots\; j_\alpha} \; \delta(\underline{r}) \tag{8.18}$$

where the coefficients $P^{(\alpha)}_{ij_1\dots j_\alpha}$ are given by

$$P^{(\alpha)}_{ij_1\dots j_\alpha} = \frac{(-1)^\alpha}{\alpha\,!} \int f_i(\underline{r}) \; x_{j_1}\dots x_{j_\alpha} \; dv \quad . \tag{8.19}$$

Since the force density must satisfy the equilibrium conditions (2.1) and (2.2) the first coefficient in (8.19) vanishes and the remaining coefficients are completely symmetric. In centrosymmetric materials tensors of odd rank would vanish. Omitting these tensors we can write

$$f_i(\underline{r}) = P^{(1)}_{ij} \, \partial_j \, \delta(\underline{r}) + P^{(3)}_{ijkl} \, \partial_j \, \partial_k \, \partial_l \, \delta(\underline{r}) + \ldots \quad (8.20)$$

or

$$|f) = |f^{(1)}) + |f^{(3)}) + \ldots \quad (8.21)$$

Finally using the weakly nonlocal expansion (1.35) for C we can obtain for G the expansion (3.25). In this way we obtain the following expansion for interaction energies W_1 and W_2

$$W_1 = -(f^{(1)} | G^o | f^{(1)}) - (f^{(1)} | G^o | f^{(3)})$$

$$+ (f^{(1)} | G^o L^{(2)} G^o | f^{(1)}) + \ldots , \quad (8.22)$$

$$W_2 = (f^{(1)} | G^o \delta L \; G^o | f^{(1)}) + (f^{(1)} | G^o \delta L \; G^o | f^{(3)})$$

$$- (f^{(1)} | G^o \delta L \; G^o L^{(2)} G^o | f^{(1)}) - \ldots . \quad (8.23)$$

In their interesting paper Hardy and Bullough have calculated W_1 for a face centered cubic lattice of the special type we considered in section 2.[12] Using the lattice statics method of Kanzaki they obtained an interaction energy W_1 which is orientation dependent, i.e. anisotropic and varies as the inverse fifth power of the separation. The same result can be obtained, as shown by the author, using the nonlocal continuum theory.[22] For this special medium corresponding to Hardy-Bullough lattice $C^{(o)}$ and $C^{(2)}$ are given by (2.31) and (2.38).

The kernels of the first two terms of the series (3.25) for G are given by

$$G_{ik}(\underline{r}-\underline{r}') = -\frac{\delta_{ik}}{4\pi C_{11}^{(o)}|\underline{r}-\underline{r}'|} \tag{8.24}$$

$$(G^o L^{(2)} G^o)_{ik}(\underline{r}-\underline{r}') = \frac{a^2}{192\pi C_{11}^{(o)}}\Big[24\pi\,\delta_{ik}\,\delta(\underline{r}-\underline{r}')$$

$$-\Big\{\delta_{ilmn}\partial_k\partial_l\partial_m\partial_n - \delta_{ijmn}\partial_j\partial_k\partial_m\partial_n$$

$$-\delta_{ik}(\partial_1^{\,4}+\partial_2^{\,4}+\partial_3^{\,4})\Big\}\,|\underline{r}-\underline{r}'|\Big]\ . \tag{8.25}$$

In the Hardy and Bullough model a cubically situated defect exerts radial forces on its twelve nearest neighbours only. Therefore, the force distribution can be written as

$$f(\underline{r}) = \sum_{n=1}^{12} f^n\ \delta(\underline{r}-\underline{r}^n)$$

which is an array of point forces f^n located at discrete points $\underline{r}^n$ relative to the array centre at $\underline{r}$. The integrals in (8.19) have to be replaced again by summation

$$P^{(\alpha)}_{ij_1\ldots j} = \frac{(-1)}{\alpha!} \int \sum_n f_i^{\,n}\, \delta(\underline{r}-\underline{r}^n)\; x_{j_1}\ldots x_{j\alpha} \; dv$$

$$= \frac{(-1)}{\alpha!} \sum_n f_i^{\,n}\, x^n_{j_1}\ldots x^n_{j\alpha} \,. \tag{8.27}$$

The force $\underline{f}^n$ is directed along $\underline{r}^n$ hence we can write

$$\underline{f}^n = K\, \underline{r}^n \,. \tag{8.28}$$

Since

$$|\underline{f}^n| = K\,|\underline{r}_n| = K\,\frac{a}{\sqrt{2}} \tag{8.29}$$

we have

$$\underline{f}^n = \frac{\sqrt{2}}{a}\,|f^n|\,\underline{r}^n \,. \tag{8.30}$$

In this way we obtain

$$P^{(1)}_{ij} = -\sum_n f_i^{\,n}\, x_j^{\,n} = -\,4\sqrt{2}\; a\,|f^n|\,\delta_{ij} \tag{8.31}$$

$$P^{(3)}_{ijkl} = -\,\frac{1}{6} \sum_n f_i^{\,n}\, x_j^{\,n}\, x_k^{\,n}\, x_l^{\,n} \,,$$

$$= -\,\frac{\sqrt{2}}{6}\, a^3\,|f^n|\,(\,\delta_{ij}\,\delta_{kl} + \delta_{ik}\,\delta_{jl} + \delta_{il}\,\delta_{jk} - \delta_{ijkl}) \tag{8.32}$$

Thus $\underline{f}^{(1)}$ and $\underline{f}^{(3)}$ are given by

$$f_i^{(1)} = -\,4\sqrt{2}\; a\,|f^n|\;\partial_i\,\delta(\underline{r}) \,, \tag{8.33}$$

$$f_i^{(3)} = -\,\frac{2}{6}\, a^3 |f^n| (\delta_{ij\;kl} + \delta_{ik}\delta_{jl} + \delta_{il}\delta_{jk} - \delta_{ijkl})\,\partial_j\partial_k\partial_l\,\delta(\underline{r}) \,. \tag{8.34}$$

Substituting these expressions as well as (8.24) and (8.25) in (8.22) and neglecting terms proportional to delta function we finally obtain

$$W_1 = -\frac{105\, a^4 |f^n|^2}{\pi C_{11}^{(o)} |\underline{r}-\underline{r}'|^5} \left[\frac{(x_1-x'_1)^4 + (x_2-x'_2)^4 + (x_3-x'_3)^4}{\underline{r}^{\,4}} - \frac{3}{5}\right] \tag{8.35}$$

which is identical to the result given by Hardy and Bullough which they obtained using the lattice theory. It is interesting to note that the term inside the brackets is actually proportional to the fourth kubic harmonic which we mentioned earlier. It is given by

$$K_4(\underline{r}) = \frac{5}{4}\left(\frac{21}{4\pi}\right)^{1/2} \left[\frac{x_1^{\,4} + x_2^{\,4} + x_3^{\,4}}{\underline{r}^{\,4}} - \frac{3}{5}\right]. \tag{8.36}$$

Hence W_1 can be written as

$$W_1 = \frac{8\sqrt{21}\; a^4 |f^n|^2}{C^{(o)} |\underline{r}-\underline{r}'|^5} K_4\,(\underline{r}-\underline{r}')\,. \tag{8.37}$$

This result could, therefore, have been obtained also by expanding in terms of kubic harmonics.

In the same way we can calculate W_2. First of all δC has to be calculated in terms of the perturbed force constants $\delta\Phi$. We first expand C in a double multipole series. Taking only the first term $\delta C^{(o)}$ it can be shown that

$$\delta C^{(o)} = \frac{N_d}{N v_o} \sum_m \sum_n \underline{r}^m \cdot T(\underline{r}^m, \underline{r}^n) \cdot \underline{r}^n \qquad (8.38)$$

where N_d is the number of defects, N is the number of atoms and T-matrix is given by[23]

$$T = \delta\Phi - \delta\Phi (\Phi + \delta\Phi)^{-1} \delta\Phi \quad . \qquad (8.39)$$

It turns out that, in the first approximation, δC has the symmetry of the defect, e.g. for a defect of cubic symmetry δC is also cubic. Taking into account this fact one finds that the first term in W_2 is proportional to the second derivative of delta function and the second term in (8.23) is proportional to $\frac{1}{|\underline{r}-\underline{r}'|^6} K_4 (\underline{r}-\underline{r}')$. So we have the interesting result that even if the medium is isotropic in the long wave length limit this interaction is in general anisotropic, in agreement with another result of Hardy and Bullough.

9. Dislocations in a continuum

A dislocation, as we mentioned in the beginning, is a type of line imperfection in an otherwise perfect crystal. It is characterized by two vectors $\underline{t}$ and $\underline{b}$. $\underline{t}$ is the unit vector defining the direction of the dislocation line and $\underline{b}$ is the displacement vector associated with the dislocation (the socalled Burgers vector). In solid state physics the Burgers vector is defined in the following way. We first form in the perfect crystal a closed circuit as shown in fig. 3a. We now draw the corresponding circuit in the dislocated crystal following precisely the same sequence of lattice vectors, as shown in fig. 3b. We assume that this

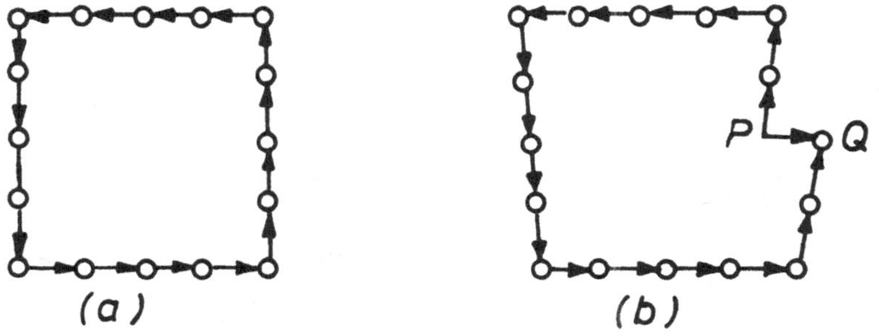

Figure 3

circuit encloses an edge dislocation. The starting point P and the end point Q are no longer coincident. The vector drawn from starting point P to the end point Q is defined as the Burgers vector which is perpendicular to $\underline{t}$. For a screw dislocation it is parallel to $\underline{t}$. We have drawn the circuits anticlockwise. Therefore, if we follow the right hand screw convention the positive sense of the unit vector $\underline{t}$ is away from the paper.

This definition can be adopted for dislocations in an elastic continuum (Volterra dislocations). A hollow circular cylinder is cut through on one side by a half-plane through the axis of the cylinder (Fig. 4a) and the two faces of the cut are given a rigid displacement relative to one another. For edge dislocations this displacement is normal to the cylinder axis (Fig. 4b) and for screw dislocations it is parallel to the cylinder axis (Fig. 4c).

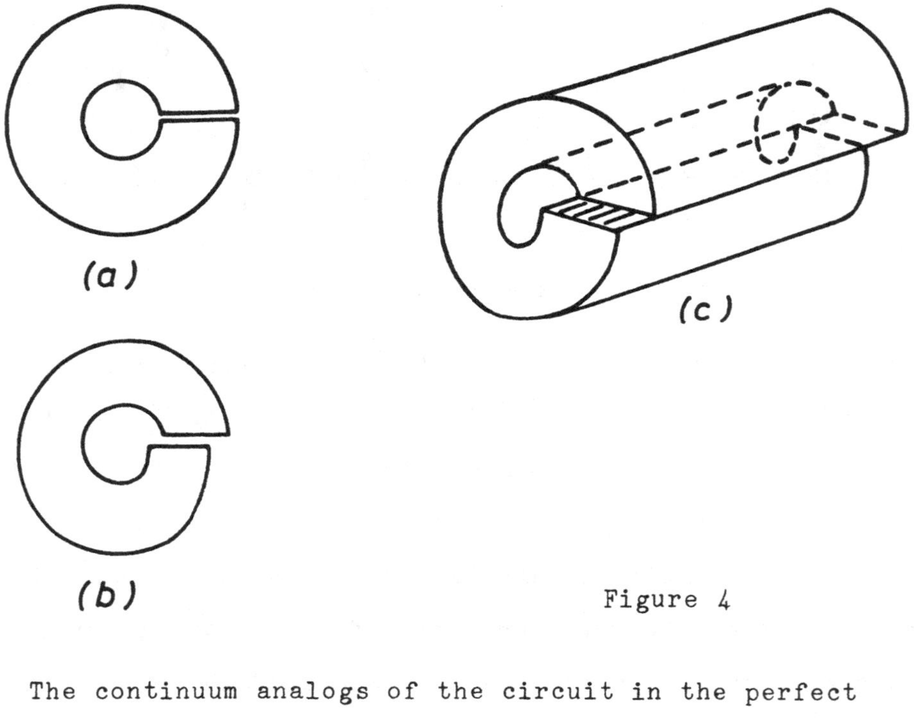

Figure 4

The continuum analogs of the circuit in the perfect crystal and the dislocated crystal are shown in fig. 5. The enclosed circuit C (Fig. 5a) in the undeformed material becomes an open circuit with non-coincident starting point P and end point Q in the deformed crystal (Fig. 5b). This

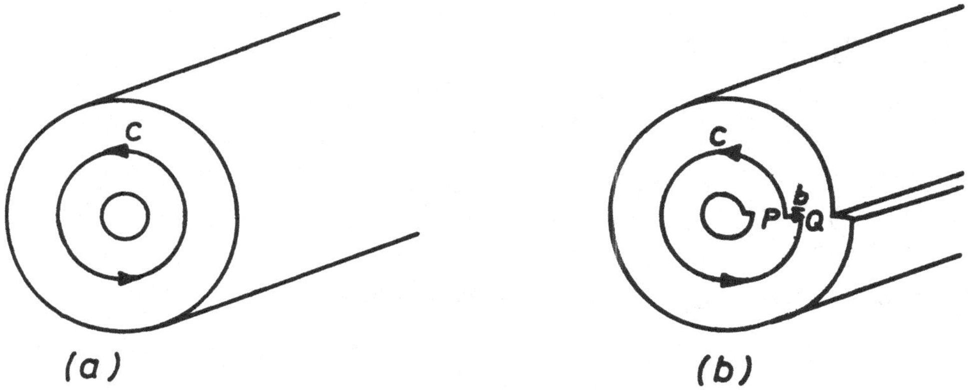

Figure 5

can be mathematically expressed as

$$\oint_C d\underline{u} + \int_Q^P d\underline{u} = \oint d\underline{u} - \int_P^Q d\underline{u} = 0 \quad . \tag{9.1}$$

It follows that

$$\underline{b} = \int_P^Q d\underline{u} = \oint_C d\underline{u} \quad . \tag{9.2}$$

In order to possess the property demanded by the above equation, the displacement field $\underline{u}$ of a dislocation must be either discontinuous or multivalued (the two points of view are equivalent). This, however, violates the St Venant's compatibility condition (4.1). That means the strains due to dislocations are incompatible and as such they are sources of internal stresses. In this case we replace the relation

$du_i = dx_j \partial_j u_i(\underline{r})$ by the Pfaffian form

$$d\, u_i(\underline{r}) = d\, x_j\, \beta^p_{ji}(\underline{r}) \tag{9.3}$$

which is not integrable because

$$\oint_C d\, x_j\, \beta^p_{ji}(\underline{r}) = b_i \neq 0\ . \tag{9.4}$$

The corresponding strains are now given by

$$\varepsilon^p_{ij}(\underline{r}) = \frac{1}{2}\left[\beta_{ij}(\underline{r}) + \beta_{ji}(\underline{r})\right]\ . \tag{9.5}$$

This is the same what we called stress-free strain in section 7.

Now let C be a circuit enclosing several dislocation lines. Since C can be considered as equivalent to a number of circuits, one around each of the lines, the right hand side of (3.4) is replaced by the sum of the Burgers vectors of all the dislocation lines threading C which we denote by $\underline{b}$. Applying Stokes' theorem to the line integral (9.4) we thus obtain

$$\Delta b_i(\underline{r}) = \int_S dS_j\, \epsilon_{jkl}\, \partial_k\, \beta^p_{li}(\underline{r})\ . \tag{9.6}$$

If the circuit C is sufficiently small we can replace the integrand in (9.6) by its value at some point on the surface and take it out from under the integral sign. The remaining integral then gives simply the area ΔS of the surface bounded by the contour C and we thus obtain

$$\Delta b_i(\underline{r}) = \Delta S_j \, \epsilon_{jkl} \, \partial_k \, \beta_{li} \quad . \tag{9.7}$$

In the limit when C is an infinitesimal circuit we can write

$$db_i(\underline{r}) = dS_j \, \epsilon_{jkl} \, \partial_k \, \beta_{li}(\underline{r}) \; . \tag{9.8}$$

From a physical point of view dislocation density can be defined in the following way. We suppose that in the neighbourhood of the point in question there are N dislocations crossing a unit area normal to $\underline{t}$. Then we define the dislocation density by

$$\alpha_{ij}(\underline{r}) = N(\underline{r}) \, t_i \, b_j \quad . \tag{9.9}$$

The number of dislocations threading the infinitesimal circuit with cap dS_j is then $dS_j \, N \, t_j$ and accordingly the total Burgers vector is

$$db_i(\underline{r}) = dS_j \, N(\underline{r}) \, t_j \, b_i = dS_j \, \alpha_{ji}(\underline{r}) \; . \tag{9.10}$$

A comparison with (9.8) shows that

$$\alpha_{ji}(\underline{r}) = \epsilon_{jkl} \, \partial_k \, \beta^p_{li}(\underline{r}) \; . \tag{9.11}$$

With this definition we see at once that the divergence of dislocation density vanishes

$$\partial_j \alpha_{ji} = 0 . \tag{9.12}$$

This result implies that the dislocation line can not end inside the crystal. Therefore, the total dislocation flux b_j through an arbitrary closed surface inside the body also vanishes

$$\int \partial_j \alpha_{ji} \, dv = \int dS_j \, \alpha_{ji} = \int db_i = b_i = 0 . \tag{9.13}$$

From our discussion in section 7 we recall that total distortion is given by

$$\beta^t_{ij} = \beta^p_{ij} + \beta^e_{ij} \tag{9.14}$$

and this distorsion is compatible, i.e.

$$\epsilon_{jkl} \partial_k \beta^t_{li} = 0 . \tag{9.15}$$

Hence we get the relation

$$\alpha_{ji} = - \epsilon_{jkl} \partial_k \beta^e_{li} . \tag{9.16}$$

Now we shall drop the superscript e and it is understood that β_{ij} is the elastic distorsion. Applying $\epsilon_{mni} \partial_n$ to both sides of (9.16) and taking the symmetric part in m and j we obtain the following relation between the dislocation density and incompability

$$(\epsilon_{mni} \partial_n \alpha_{ji})_{(mj)} = - \epsilon_{mni} \epsilon_{jkl} \partial_n \partial_k \varepsilon_{li} = - \eta_{mj} . \tag{9.17}$$

In the case of a single dislocation we have[10, 24]

$$\alpha_{ji}(\underline{r}) = \delta_L(\underline{r})\; t_j\; b_i \quad . \tag{9.18}$$

The distribution $\delta_L(\underline{r})$ is defined by

$$\delta_L(\underline{r}) = \int_L \delta_B\,(\underline{r}-\underline{r}_L)\; dL \tag{9.19}$$

where $\underline{r}_L$ is on the dislocation line. It follows that

$$\int \delta_L(\underline{r}')\; f(\underline{r}-\underline{r}')\; dv = \int_L dL \int \delta_B(\underline{r}'-\underline{r}_L)\; f(\underline{r}-\underline{r}')\; dv'$$

$$= \int_L dL\; f(\underline{r}-\underline{r}_L) \quad , \tag{9.20}$$

$$\int (\nabla'\delta_L(\underline{r}'))\; f(\underline{r}-\underline{r}')\; dv' = \int dL\;\; \nabla f(\underline{r}-\underline{r}_L) \;, \tag{9.21}$$

$$\int \delta_L(\underline{r})\; dS = 1 \quad . \tag{9.22}$$

$\delta_L(\underline{r})$ is thus a two-dimensional distribution centered around the dislocation line. Suppose now we take the surface perpendicular to the dislocation line then we can write $dS_j = t_j\, dS$ and so we have

$$\int_S dS_j \quad {}_{ji}\,(\underline{r}) = \int_S \delta_L(\underline{r})\; b_i\; t_j\; t_j\; dS \; . \tag{9.23}$$

Since $t_j\, t_j = 1$ we get in view of (9.22)

$$\int_S dS_j \quad {}_{ji}\,(\underline{r}) = b_i \tag{9.24}$$

as it should be.

10. Interaction energy of dislocations

We now consider the interaction between two dislocations which can be considered as two internal stress systems. According to the results of section 6 the interaction between two internal stress systems is given by

$$W_{int} = (\eta^A \mid J \mid \eta^B) . \tag{10.1}$$

Using the relation (9.17) we can express the interaction energy in terms of dislocation density. However, we first observe that $J_{ijkl}(\underline{r}-\underline{r}')$ is symmetric in the indices i and j as well as k and l. In fact all modified Green operators have the symmetry of C operator. Therefore, the interaction energy can be written as

$$W_{int} = (\alpha^A \mid \text{curl } J \text{ curl} \mid \alpha^B) . \tag{10.2}$$

In view of the relation (9.18) and (9.20) we can write it in the form

$$W_{int} = b_k^A \, M_{kp}^{AB} \, b_p^B \tag{10.3}$$

where

$$M_{kp}^{AB} = \int_{L^A} \int_{L^B} \epsilon_{ijk} \, \epsilon_{mnp} \, \partial_j \, \partial_n \, J_{ilqm}(\underline{r}_L{}^A - \underline{r}_L{}^B) \, dL_l^A \, dL_q^B . \tag{10.4}$$

This quantity is sometimes called the dislocation mutual

inductance because of the close similarity between the interaction energy (10.3) of two dislocation loops and the corresponding result for current loops in magnetostatics. In the local theory interaction energy in this form was first given by Kröner.[19]

For the self energy of a dislocation equation (10.3) is still valid provided we replace B by A and multiply by $\frac{1}{2}$, i.e.

$$W_{self} = \frac{1}{2} b_k^A M_{kp}^{AA} b_p^A \quad . \tag{10.5}$$

As one can see from (10.4) one can calculate M_{kp}^{AA} by integrating twice over the same curve L^A. In the local theory the integrand is singular and the result diverges. However, this is not the case in the nonlocal theory.

Combining (10.3) and (10.5) we can write the energy of a number of dislocations in the form

$$W_{int} = \frac{1}{2} \sum_{A,B} b_k^A M_{kp}^{AB} b_l^B \quad . \tag{10.6}$$

For further discussion it is convenient to introduce the concept of interaction energy E per unit length of dislocations. In the local theory it is usually defined in the following way. One assumes that two dislocations have finite lengths. One then performs the integration in (10.4) and divides the result by dislocation length. Here we define E as

$$E = b_k^k \, T_{kp}^{AB} \, (\underline{R}_L) \, b_p^B \tag{10.7}$$

where

$$\underline{R}_L = \underline{r}_L A - \underline{r}_L B \; , \tag{10.8}$$

$$T_{kp}^{AB} = \epsilon_{ijk} \, \epsilon_{mnp} \, \partial_j \, \partial_n \, J_{ilqm}(\underline{R}_L) \, t_l^A \, t_q^B \; . \tag{10.9}$$

For the sake of simplicity we can now consider straight dislocations only. We assume that dislocation lines are parallel to x_3-axis. Hence

$$T_{kp}^{AB} = \epsilon_{ijk} \, \epsilon_{mnp} \, \partial_j \, \partial_n \, J_{i33m}(\underline{R}_L) \tag{10.10}$$

where we have used the fact that $t_3^A \, t_3^B = 1$ because $\underline{t}$ is a unit vector.

Consider now two screw dislocations along the x_3-direction. The only nonvanishing components of the Burgers vectors are along the x_3 direction. Hence putting $b_3^A = b^A$ and $b_3^B = b^B$ we get

$$E_{int} = b^A \, b^B \, \epsilon_{ij3} \, \epsilon_{mn3} \, \partial_j \, \partial_n \, J_{i33m}(\underline{R}_L) \; . \tag{10.11}$$

If we now neglect the inhomogeneity effect and assume that the medium is an isotropic Debye continuum we use the expression (5.30) for J with the only difference that g_B is given by the two-dimensional formula (1.34). We then obtain

$$E_{int} = b^A\, b^B \mu \left(\frac{\partial}{\partial x_1{}^2} + \frac{\partial}{\partial x_2{}^2}\right) g_B\, (\underline{R}_L)$$

$$= b^A\, b^B\, \mu \Delta\, g_B(R_L) = b^A\, b^B\, G_B(\underline{R}_L) \qquad (10.12)$$

where G_B is now given by (1.33). It depends on $R = |\underline{R}_L|$ as well as on L. One can see that interaction energy given by (10.12) diverges as $L \rightarrow \infty$. However, we can argue that L is never infinite since all crystals are finite. Physically this divergence indicates how important the elastic energy of the dislocations is as a part of the free energy.

In the limit when the Brillouin zone radius becomes very large or the distance $R = |\underline{R}_L|$ is very large ($k_B R >> 1$) the function G_B tends to the usual Green function of the Laplacian in two dimensions. Thus we obtain the local approximation

$$E_{int} = \frac{\mu}{2\pi}\, b^A\, b^B \ln \frac{L}{R}\ . \qquad (10.13)$$

Even more interesting is the case of self energy which can be interpreted in the following sense. If there exists a dislocation in the medium the energy of the medium is higher than it would be if there were no dislocations. The difference between the two energies is the self energy. In the local theory one has to use rather involved arguments to find an expression for the self energy. Here it can be given in a simple way by

$$E_{self} = \frac{1}{2} b^2 \mu \; G_B(0) = \frac{b^2\mu}{4\pi} \ln k_B \; L \qquad (10.14)$$

and this is finite whereas in the local theory one would get a divergent expression.

It is customary, by analogy with general procedures in mechanics, to introduce the notion of an interaction force per unit length on a dislocation line, which is minus the rate of change of energy E with respect to the dislocation separation. This force can be obtained from (10.12) as

$$f = -\frac{d}{dR} E_{int}(R) = \frac{\mu}{2\pi R} \; b^A \; b^B \; (1 - J_o(k_B R)) \, . \qquad (10.15)$$

That means this force has an oscillatory character.

Let us now consider two edge dislocations along x_3-axis. Their Burgers vector lie in the plane perpendicular to x_3-axis. Let these Burgers vectors be in the x_1 direction or one may be in x_1 direction and the other opposite to it. We then have

$$E_{int} = \pm \, b^A \; b^B \, \epsilon_{ij1} \, \epsilon_{mn1} \, \partial_j \, \partial_n \, J_{i33m}$$

$$= \pm \frac{\partial^2}{\partial x_2^{\,2}} \, g_B \; (R_L) \qquad (10.16)$$

and the force along the slip plane is

$$f_1 = \mp \frac{4\mu(\lambda + \mu)}{\lambda + 2\mu} \; \frac{\partial}{\partial x_1} \; \frac{\partial}{\partial x_2^{\,2}} \; g_B(R_L) \, . \qquad (10.17)$$

In the local approximation it redues to the form

$$f_1 = \mp \frac{\mu(\lambda+\mu)}{\pi R(\lambda+2\mu)} \cos\theta \cos 2\theta \,. \tag{10.18}$$

Similar results have been obtained by Vdovin and Kunin by a somewhat different method.[10]

If one is edge and the other is screw dislocation along the x_3-axis with Burgers vectors respectively along x_1 and x_3 axis we get

$$E_{int} = b_1^A\, b_3^B\, \epsilon_{ij1}\, \epsilon_{mn3}\, \partial_j\, \partial_n\, J_{i33m} = 0 \,. \tag{10.19}$$

The inhomogenity effect associated with the dislocation core can also be considered in the present formalism but it is a bit more complicated because of the much lower symmetry of the problem. On the two sides of the slip plane the change in the force constants is of very different type because on one side the lattice is compressed due to the presence of an extra half-plane of atoms whereas on the other side the lattice is dilated. However, two qualitative conclusions can be drawn from the T-matrix formulation. Firstly the screw dislocations will interact with the edge dislocations. Secondly the self-energy of a screw dislocation is not altered by the inhomogeneity.

Acknowledgement

The author would like to thank Professor E. Kröner for helpful suggestions.

References

1. Rogula, D., Bull. Acad. Polon. Sci., Ser. Sci. Techn. 13, 7 (1965)
2. Krumhansl, J.A., in: Lattice Dynamics, Ed. R.F. Wallis, Pergamon Press, London, Oxford 1965 (p. 627)
3. Kröner, E. and B.K. Datta, Z. Phys. 196, 203 (1966)
4. Kunin, I.A., Prikl. Mat. Mekh. 30, 642 (1966)
5. Eringen, A.C. and D.G.B. Edelen, Internat. J. engng. Sci. 10, 233 (1972)
6. Shannon, C.E., Proc. IRE, January 1949
7. Gairola, B.K.D., Arch. Mech. 28, 393 (1979)
8. Kotowski, R., Z. Phys. B 33, 321 (1979)
9. Kosilova, V.G., I.A. Kunin, and E.G. Sosnina, Fiz. Tverd. Tela, 10, 367 (1968)
10. Vdovin, V.E. and I.A. Kunin, Fiz. Tverd. Tela, 10, 375 (1968)
11. von der Lage, F.C. and H.A. Bethe, Phys. Rev. 71, 612, (1947)
12. Gairola, B.K.D., Nonlocal Theories of Material systems, Jablonna 1975, Ossolineum 1976
13. Hardy, J.R. and R. Bullough, Phil. Mag. 15, 237 (1967), ibid 16, 405 (1967)
14. Kröner, E., Z. Phys. 142, 463 (1955)
15. Kunin, I.A., PMM 31, 889 (1967)
16. Kröner, E., to be published in Journal of the Engineering Mechanics Division of American Society of Civil Engineers

17. Dederichs, P.H. and R. Zeller, KFA-Jül-Report, Jül-877-FF, 1972

18. Kröner, E., Z. Phys. 139, 175 (1954), ibid 143, 374 (1955)

19. Kröner, E., Kontinuumstheorie der Versetzungen und Eigenspannungen. Ergeb. angew. Math. 5 (1958)

20. Marguerre, K., ZAMM 35, 242 (1955)

21. Colonetti, G., Atti Acad. naz. Lincei Re 27/2, 155 (1918)

22. Gairola, B.K.D., phys. stat. sol. (b) 85, 577 (1978)

23. Dederichs, P.H. and C. Lehmann, Z. Phys. B 20, 155 (1975)

INTRODUCTION TO NONLOCAL THEORY OF MATERIAL MEDIA

DOMINIK ROGULA

Laboratoire de Mécanique Appliquée, Besancon
Institute of Fundamental Technical Research, Warsaw

I. VARIOUS CONCEPTS OF NON-LOCALITY

1. Nonlocal interactions in nature

One of crucial concepts of natural science is that of interaction between various objects in the world.

It is this interaction that makes our world so interesting to observe and so difficult to understand for a scientist or a philosopher. We shall focus our attention on a certain aspect of the spatial range of interactions between material objects. The interactions that occur only when the interacting objects touch each other will be referred to as local interactions. In the case of interactions which can occur when the interacting objects are separated in the physical space there are, in principle, two possibilities. Either such interaction can be explained in terms of a sort of infinite chaining or propagation of some local interactions, or not. If not, i.e. if the distant interaction is considered irreducible, or even, within the framework of a particular theory, not considered reducible to local ones, then it will be referred to as nonlocal interaction.

Is the real physical world local in the above defined

sense, or not?

The knowledge gained by man from first physical experience suggested the answer to the above question in the affirmative. Although from the very beginning of the history of mankind the existence of nonlocal interactions of magic nature was strongly believed, the physical interactions appeared rather local. In order to make an object move, it was necessary to touch it. Aristotle's statement "It is evident, that between the extremities of the moved and the movent, that are respectively first and last in reference to the moved there is nothing intermediate" can be understood as, in our languag , the locality principle of causal interactions. The phenomenon of ἤλεκτρον seemed peripheric, outside the main course of natural events.

For the first time in the history of human knowledge non--locality occured in a serious manner in the Newtonian law of gravitation (1687) which implies existence of interaction between bodies distant from one another. The weight of a body so far considered self-contained turned out to be a result of nonlocal interaction with another body, the earth. Almost a century later (1785) Coulomb demostrated validity of a similar law for electric forces.

Those and many other instances of nonlocal laws of interaction were, however, wondered at. And not only wondered:

in about the middle of the nineteenth century Faraday and Maxwell totally explained the interactions between electric charges and/or currents in terms of local interactions carried by electromagnetic field. The discovery of electromagnetic waves showed that the field is not just an auxilliary notion barely convenient in description of interactions at distance, but that it is a physical entity which can be separated from, and can exist independently of its sources. The concept of physical field was also succesful in Einstein's theory of gravitation (1915) which corrected the Newtonian one, while the relativistic kinematics of the space-time put non-locality in strong contradiction with causality. The physical world reappeared local.

It was, however, a temporary impression. All the efforts to create a coherent physical theory based on local fields, classical or quantum, failed in the domain of micro-physics. The reason was the singularities of the type of ultra-violet catastrophe, thus connected with small distances. They result in divergence of theoretical expressions for almost all physical quantities. It turned out that logically coherent quantum field theory being relativistic and causal excludes any interaction between particles. In other words, the only truly local interactions equal identically null.

The verdict is not final as the subsequent development of the theory of elementary particles has not brought forth any definite solution.

The contradiction between relativity and non-locality restricts occurence of the latter in contemporary physics. A non-local theory can, first of all, be conceived as an approximation of a more exact, and more complicated, relativistic theory, and considered applicable whenever relativistic effects are small. On the other hand, a non-local theory may be constructed with the aim to provide a better description of matter at the sub-micro-level. In the last case, it is the theory of relativity that would be less exact. It should however follow from such refined non-local theory as an approximation valid for sufficiently great distances.

2. Nonlocal interactions in material media

At present there are two widely accepted conceptual models of material media: the material continuum, where the matter is considered spread in a continuous way over a certain region, and the molecular model, where the matter is regarded as composed of separate molecules in motion. The latter is physically deeper, while the former is simpler. From the macroscopic point of view the continnum model represents more rough approximation of physical reality than the molecular one, and many papers have been devoted to the problem of derivation or justification of the material continuum in terms of the molecular model. It is however worthwhile to stress that conceptually those models are independent and that, for instance, one may apply the idea of continuum not only at the usual macro-level but also at the sub-micro-level by considering the elementary particles as having deeper intrinsic structure of (quantum) continuum.

An immediate consequence of the spatial separation of single molecules of material body is that, excepting collisions, there are no contact interactions between them. Thus, in a theoretical model, for getting any interaction between distant molecules, two principal possibilities are open: either one adds, as an extra component of the material body, some physical fields which would serve as carriers of local interactions between molecules, or one is forced to introduce some non-local interaction. If one chooses the first possibility, one obtains a complicated theory with retardation of interactions etc., which can only be pragmatically justified if the corresponding relativistic effects are important. If, as in most cases, these effects may be disregarded, the molecular model with appropriate nonlocal interactions is used. The idea of such models is historically earlier than that of relativity. A great advantage of models with instantaneous nonlocal interactions is that they are much simpler than the corresponding relativistic models would be. Such simplification can be obtained from a relativistic theory by considering its limit for the light velocity $c \to \infty$.

While occurence of nonlocal interaction in the molecular model of matter appeared quite natural, if not indispensable, the contrary can be said concerning the material continuum. As conceived by Cauchy (1823), classical continuum mechanics was based on contact interaction forces resulting from the corresponding stress tensor. The contact forces were assumed locally determined in the following sense: at a given point $\underline{x}$ of the material body the stress tensor depends only on infinitesimal enviroment of this point, excluding any dependence on all the points $\underline{x}'$ placed at finite distance from $\underline{x}$. Those assumptions were never believed, even by Cauchy himself, to express the deepest physical truth. They, however, allowed to create a simple and powerful continuum mechanics, which turned out to be a great success of science.

The idea of locally determined contact forces was, for almost 150 years, considered a necessary ingredient of continuum mechanics. It was not until 1960's when it was clearly recognized that the concepts of material continuum and the range of interaction forces are independent of each other, and when first papers on nonlocal continuum theory appeared, starting from the paper by Kröner and Datta (1966) which derived a nonlocal continuum as an approximation of crystal lattice. The idea of nonlocal continuum has also been formulated and developed by other authors, particularly by Kunin, Edelen, Eringen and the lecturers of the present course.

How such an idea can be physically justified? To get an answer to this question it is necessary to compare nonlocal continuum theory with both classical continuum and the molecular theories. It is clear that, in spite of its power in its domain of application, the classical continuum theory fails to describe many important phenomena. Let us quote Kröner (1969):

"One finds that the conventional elasticity theory is good as long as the true solution of a problem, when given in terms of Fourier integrals and sums, involves, with essential amplitudes, only such wavelengths which are large compared to the range of cohesive forces. This is the wavelength interval in which dispersion is negligible.

On the other hand, it is clear that in order to describe phenomena on an atomic scale no theory can be used which overlooks discreteness of the matter. In other words: the non-local continuum theory which extends the conventional elasticity theory towards shorter wavelength ceases to apply where the wavelengths become comparable with the lattice parameter. Here a lattice theory must be applied."

Thus, if one considers a material whose typical intermolecular distance is $\underline{a}$ and typical range of intermolecular forces is $\underline{l}$, then for characteristic distances λ such that

$$a \ll \lambda \ll l \qquad (2.1)$$

the local theory is no longer valid, while application of the molecular model is not yet necessary. In this sense, inequality (2.1) defines a natural domain of nonlocal continuum theory. Evidently, for existence of such a domain it is necessary that $\underline{l}$ be much greater than $\underline{a}$,

$$a \ll l \ , \qquad (2.2)$$

but such a relation may be expected to hold in many real materials.

We have, however, not yet proven that there is a real need for nonlocal continuum. One may argue at this point that there is no such need, since inequality (2.1) only admits continuum but does not exclude molecular lattice as representation of matter and, in consequence, molecular theory can be used in this domain. This argument is a perfectly valid one. The nonlocal continuum theory can be fully justified only if it produces some models which would be able to give physically valid results in the domain (2.1) but in a much simpler way than in molecular theory. The example of classical continuum theory is, however, encouraging here: it has shown that the idea of continuum is able to bring forth a great theoretical simplification without losing the validity of results in an appropriately chosen domain.

3. Mathematics of non-locality

Consider a physical theory based on a fundamental equation of the form

$$Ay = z \tag{3.1}$$

The exact physical meaning of this equation need not be precised here. It can play the role of a governing equation of the theory, or a basic constitutive relation, or of something else. As an example we can consider elasticity with y - displacements, z - external forces, or with y - deformations, z - stresses, or electrostatics with y - electric field, z - free charge, or y - electric field, z - electric induction, etc..

From the mathematical point of view this equation will be considered as defining the action of a certain operator A. We assume that $y = y(x)$ and $z = z(x)$ are functions or distributions over a certain region Ω in (Euclidean) space. Generally, y and z will be considered elements of some well defined spaces Y and Z, respectively. Strictly speaking those spaces need not necessarily be composed of functions or distributions in the classical sense: it is sufficient that the concept of <u>support</u> with its usual properties be defined for $y \in Y$, $z \in Z$.

In classical theories the oparator A is usually local. Colloquially, it means that equation (3.1) states no direct relation between $y(x)$ and $z(x')$ for $x \neq x'$. This idea can be expressed in a precise way with the aid of the concept of support.

Definition (3.1) . The operator A in equation (3.1) is called local if

$$\text{supp} \quad Ay \subset \text{supp } y \tag{3.2}$$

for each $y \in Y$, and

$$\text{supp } (Ay_2 - Ay_1) \subset \text{supp } (y_2 - y_1) \tag{3.3}$$

for every $y_1, y_2 \in Y$.

The operator A can, in general be nonlinear. If it is linear then condition (3.2) alone is sufficient in the above definition. For a nonlinear operator, condition (3.3) is necessary; in this case, instead of (3.2) one can require that

$$y = 0 \longrightarrow Ay = 0 \tag{3.4}$$

Definition (3.1) expresses mathematically the basic idea of non-locality. More refined questions, for which some knowledge of nonlocal theory is already useful, will be discussed in the sequel.

4. Weak non-locality

Now we shall discuss briefly an alternative definition of (non)locality of material media. Roughly speaking, one may say that a material medium is nonlocal in this new sense, if its physical properties determine an intrinsic measure of length. Or, in other words, the medium is local in the new sense if the coefficients that fully characterize its physical properties cannot be combined into a parameter of the dimension of length.

This new concept will be referred to as weak non-locality. It can be defined more precisely in the following way.

Consider a scale transformation

$$\underline{x} \rightarrow \underline{x}' = a^{-1}\,\underline{x} \qquad (4.1)$$

of the physical space. Let S be the transformation of spaces Y and Z resulting from this scale transformation (for simplicity we use the same symbol S for both spaces Y, Z):

$$y \rightarrow y' = Sy\ ,$$

$$z \rightarrow z' = Sz\ . \qquad (4.2)$$

Now we can give the following

<u>Definition (4.1)</u>. The operator A in (3.1) is called <u>strictly local</u> if it is local in the sense of Definition (3.1) and if the equation (3.1) is invariant with respect to the scale transformation (4.1), (4.2) . An operator which is not strictly local will be called <u>weakly nonlocal</u>.

The invariance condition required in the above definition can be written as

$$A = S^{-1}\,AS\ , \qquad (4.3)$$

or equivalently

$$SA = AS \qquad (4.4)$$

It follows directly from the definitions that

$$\text{strictly local} \Longrightarrow \text{local} ,$$
$$\text{nonlocal} \Longrightarrow \text{weakly nonlocal.} \qquad (4.5)$$

The reader may verify that the classical theory of elasticity is strictly local, while the strain gradient theory, discussed in Chapter IV is weakly nonlocal. The integral models discussed in Chapter II are nonlocal in the strong sense.

5. Alocal and strictly nonlocal theories

As the question of locality or non-locality is closely related to the physical and mathematical foundations of the theory of material continua in general, the problems involved can be subdivided into two classes. The first of them is concerned with the relevant aspects of general continuum theory that are independent of any specific assumptions of locality or non-locality. Those are the matters which in the logical order precede such assumptions, constituting the alocal part of the theory. The other part is concerned with definitely nonlocal problems arising in connection with rejection of Cauchy's or other local postulates. In a free analogy if one compares the local postulates to the Euclidean axiom of paralels, the above two parts of the theory can be compared to the absolute and to the non-Euclidean geometry, respectively.

II. SIMPLE NONLOCAL MODELS

1. Idea. Basic relations

We shall begin our study of nonlocal theory of material media from construction of models which in a simplest way express the idea of non-locality in continuum mechanics. The basic heuristic principle in construction of such models consists in taking, in place of the classical local relations, some integral relations.

As a point of departure we shall take an equation expressing the balance of momentum in a continuous medium which (under classical assumptions concerning continuity of the medium and that of distribution of physical quantities) can be written in the form

$$\dot{p}_i + (p_i v_k)_{,k} + r_i = f_i \tag{1.1}$$

where the symbols employed have the following meaning. The density of linear momentum has been denoted by p_i, the velocity of a material point by v_i, and the density of external force, i.e., the force with which the considered medium is acted upon by the remaining world, by f_i . The symbol r_i stands for the density of forces with which a material particle acts on the surrounding medium, and which arise as a result of the interaction between material particles of the body. All these quantities depend, in general, on the space co-ordinates $\underline{x}$ and the time t. The dot over a symbol denotes the differentiation with respect to t (the geometrical co-ordinates $\underline{x}$ fixed!) while the comma preceding, say, the index k, introduces the partial differentiation with respect to x_k .

The quantities p_i, v_i, f_i in nonlocal theory have their classical meaning and are not modified. The forces of interaction between particles of the medium enter the equation (1.1) through r_i , and this quantity is of our present concern.

In an elastic medium the forces r_i at the same time t are determined by the state of deformation of the medium at the time t and, in consequence, by the displacement field at the same time. Hence, there is an equal-time functional dependence.

$$r_i = r_i[u_j] \tag{1.2}$$

between the fields r_i and u_j .

Thus, in considerations concerning the elastic media there is no need to take the time into account an explicit way. It is sufficient to treat it as an implicit parameter. Moreover, without losing anything essential one can confine himself to statics, since combined with equation (1.1) the static relation (1.2) determines the dynamical properties of the medium.

By substituting $p_i = 0$ into (1.1) one obtains the static relation for elastic media,

$$r_i[u_j] = f_i \tag{1.3}$$

which, for given external forces f_i, is to be solved to determine u_j. In classical elasticity, under assumptions of geometrical and physical linearity, but with admitting the possibility of anisotropy and inhomogeneity, the relation (1.2) has the form

$$r_i(x) = -\partial_j c_{ijkl}(\underline{x})\ \partial_i u_k(\underline{x}) \tag{1.4}$$

where $c_{ijkl}(\underline{x})$ represents the tensor of elastic modulae. It is a local relation. By replacing it by simple nonlocal relations we obtain simple models of nonlocal elastic continuum.

2. Integral model

If in place of the classical expression (1.4) a simple linear relation with an integral operator is assumed, then the equilibrium equation (1.3) can be written in the form

$$f_i(\underline{x}) = \int_\Omega dv' \Phi_{ij}(\underline{x}, \underline{x}')u_j(\underline{x}') + A_{ij}(\underline{x})u_j(\underline{x}) \tag{2.1}$$

where Ω stands for the region occupied by the body, and $\Phi_{ij}(\underline{x}, \underline{x}')$ and $A_{ij}(\underline{x})$ are certain functions. With given Ω the functions Φ_{ij} and A_{ij}, and the external forces, one may, in principle, try to solve the equation (2.1) in order to obtain the displacement field u_i .

The model just constructed, which for simplicity will be called the integral model, has to be submitted to the physically necessary invariance principles. First of all, one should require that rigid motions of the whole body produce no forces r_i .

Consider first the case of rigid translations

$$u_i = c_i = \text{const.} \qquad (2.2)$$

One obtains the following relation

$$0 = \int_\Omega dv' \; \Phi_{ij}(\underline{x}, \underline{x}')c_j + A_{ij}(\underline{x})c_j$$

which is to be valid for an arbitrary vector c_j. In consequence the equation

$$A_{ij}(\underline{x}) = - \int_\Omega dv' \; \Phi_{ij}(\underline{x}, \underline{x}') \qquad (2.3)$$

must hold. To define an integral model it is, therefore, sufficient to give the kernel $\Phi_{ij}(\underline{x}, \underline{x}')$.

This kernel cannot be yet given arbitrarily. A reasoning similar to the above but applied to infinitesimal rotations yields the condition of rotational invariance, which will not be discussed here in detail. Such a model must also conform to some other general principles which we shall formulate in Chapter II.

Now, let us note that with taking (2.3) into account, the equation (2.1) can be written as

$$f_i(\underline{x}) = \int_{\Omega} dv' \; \Phi_{ij}(\underline{x}, \underline{x}')[u_j(\underline{x}') - u_j(\underline{x})] \qquad (2.4)$$

If one compares it with the corresponding equation of discrete lattice dynamics

$$f_i^A = \sum_{B\varepsilon\Omega} \Phi_{ij}^{AB} [u_j^B - u_j^A] \; , \qquad (2.5)$$

where A, B are indices running over the material points of the lattice and Φ_{ij}^{AB} with A ≠ B are the corresponding force constants, one can immediately see the analogy. Therefore the integral model may be looked upon as a "continuous crystal".

3. Nonlocally determined stresses

Another simple model of nonlocal elastic medium can be obtained without giving up the idea of contact forces and Cauchy stress tensor σ_{ij}. In this case the classical expression of r_i by σ_{ij}, i.e.

$$r_i(\underline{x}) = - \sigma_{ij,j}(\underline{x}) \qquad (3.1)$$

is accepted, but the classical constitutive relation

$$\sigma_{ij}(\underline{x}) = c_{ijkl}(\underline{x}) \; \varepsilon_{kl}(\underline{x}) \qquad (3.2)$$

is rejected. Instead of (3.2) one again assumes a simple linear combination with integral operator

$$\sigma_{ij}(\underline{x}) = \int_\Omega dv' \Phi_{ijkl}(\underline{x}, \underline{x}') \varepsilon_{kl}(\underline{x}') +$$

$$+ D_{ijkl}(\underline{x}) \ \varepsilon_{kl}(\underline{x}) \tag{3.3}$$

with certain tensor functions $\Phi_{ijkl}(\underline{x}, \underline{x}')$ and $D_{ijkl}(\underline{x})$ which are symmetric at least in the indices k, l. These functions, similarly as in the integral case, should be submitted to appropriate conditions in order to get a model conforming to the general principles.

By making use of the above mentioned symmetry and of the expression

$$\varepsilon_{kl} = \frac{1}{2}(u_{k,l} + u_{l,k}) , \tag{3.4}$$

one arrives at the equation of equilibrium in the form

$$f_i(\underline{x}) = - \partial_j \int dv \Phi_{ijkl}(\underline{x}, \underline{x}') \ \partial_l' u_k(\underline{x}') +$$

$$+ \ \partial_j D_{ijkl} \partial_l u_k(\underline{x}) \quad , \tag{3.5}$$

where ∂_l' stands for differentiation with respect to x_l' . Thus, in the present case, one has to solve an integro-differential equation.

It is however not the only difference with respect to the integral model. According to the physical meaning of the stress tensor σ_{ij} , the boundary conditions of the form

$$q_i(\underline{x}) = - \ \sigma_{ik}(\underline{x}) n_k(\underline{x}) \ , \ \underline{x} \in \partial\Omega \tag{3.6}$$

have to be satisfied. If $q_i = 0$, i.e. if there are no

surface loads, the corresponding homogeneous boundary conditions have to be satisfied, while in the integral model there are no specific boundary conditions at all.

4. A combined model

In the models discussed so far one has totally denied at least one of the classical assumptions. There is, however, still another possibility. One can well imagine, that in a certain material medium there are double interactions: of a very short, and of a rather long range, respectively. One may be satisfied with modeling the short-range part by usual contact interactions, and try to obtain better modeling for the long-range part only. In this way one can arrive at the idea of a combined model.

The simplest combination of this kind consists in accepting the expressions (3.1) and (3.2) for modeling the short-range part, and in applying the idea of the integral model for the remaining part of interactions. After performing that procedure one obtains the equilibrium equation of the form

$$f_i(\underline{x}) = -\sigma_{ij,j}(\underline{x}) + \int_\Omega dv' \Phi_{ij}(\underline{x}, \underline{x}')[u_j(\underline{x}) - u_j(\underline{x}')] \quad (4.1)$$

with appropriate boundary conditions for σ_{ij} .

Other combinations are also, in principle, possible.

5. A troublesome integral equation

In this section we shall present an example of an apparently reasonable nonlocal model which, however, has bad properties where the solutions are concerned.

Consider an infinite medium governed by an integral equation of the form (2.1) with the kernel

$$\Phi_{ij}(x, x') = c_{iljm}\partial_l\partial_m' \,\Phi(x - x') \quad , \qquad (5.1)$$

where c_{iljm} is the classical isotropic tensor,

$$c_{iljm} = \mu(\delta_{ij}\delta_{il} + \delta_{im}\delta_{il}) + \lambda\delta_{il}\delta_{jm} \; , \qquad (5.2)$$

and the function $\Phi(x - x')$ is given by

$$\Phi = \left(\frac{\beta}{\sqrt{\pi}}\right)^3 e^{-\beta^2(x - x')^2} \; . \qquad (5.3)$$

We can also put this equation in the form (3.3) with

$$\Phi_{ijkl}(x, x') = c_{ijkl}\Phi\,(x - x') \qquad (5.4)$$

For this form of Φ , the parameter β^{-1} can be interpreted as the range of interactions. The numerical factor in (5.2) is chosen so that when $\beta^{-1} \to 0$, then

$$\Phi(x - x') \to \delta^{(3)}(x - x') \quad , \qquad (5.5)$$

and the corresponding equations become purely classical.

Now, let us try to find the fundamental solution $G_{ij}(x)$:

$$\int d^3x \;\Phi_{ij}(x - x')G_{jn}(x) = \delta^{(3)}(x)\,\delta_{in} \; . \qquad (5.6)$$

Making use of the Fourier transformation, we obtain the equation

$$\hat{\Phi}_{ij}(k)\hat{G}_{jn}(k) = \quad {}_{in} \tag{5.7}$$

for the corresponding Fourier transforms. The Fourier transform of $_{ij}$ can easily be calculated,

$$\hat{\Phi}_{ij}(k) = \int d^3x e^{-ikx} \; \Phi_{ij}(x) = [\mu k^2 \delta_{ij} + (\lambda+\mu) k_i k_j] e^{-k^2/4\beta^2} \tag{5.8}$$

and from the eqn. (5.7) we obtain:

$$\hat{G}_{ij}(k) = \frac{1}{\mu}\left(\frac{\delta_{ij}}{k^2} - \frac{k_i k_j}{k^4}\right) + \frac{1}{\lambda+2\mu}\frac{k_i k_j}{k^4} \quad e^{k^2/4\beta^2} . \tag{5.9}$$

For $1/\beta = 0$, this is the classical result. Otherwise, however, $\hat{G}_{ij}(k)$ has an exponential growth at infinity and cannot be transformed in a usual way. Therefore, we have to conclude that, in the example considered, a Fourier-transformable fundamental solution does not exist.

From the mathematical point of view, the meaning of the last statement is not quite clear. It can be made precise in terms of tempered distributions. The tempered distributions are defined as continuous linear forms on the space S which consists of infinitely differentiable functions $\ell(x)$, such that

$$\sup_x |x^\mu \partial^\nu \ell| < \infty \tag{5.10}$$

for any two multi-indices ∂ ,ν . It can easily be proved that, if $\ell \in S$ is a tempered distribution, then the convolution of ℓ and u exists, and is again a tempered distribution. Moreover, the Fourier transform of this convolution equals the product of the Fourier transforms $\hat{\ell}$ and $\hat{u}$, and is a tempered distribution, too. Thus by observing that the kernel defined by the eqns (5.1) - (5.3) belongs to the space S, we see that the eqn (5.1) becomes meaningful in the sense of convolution for any u_i which is a tempered distribution. In this case, the Fourier transformation method we have just applied to solve this equation is entirely justified. The result (5.9), being itself no tempered distribution, cannot be retransformed into a tempered distribution. Therefore, the rigorous conclusion is that there is no tempered distribution which, in the example considered, could serve as a fundamental solution.

Intuitively, the class of tempered distributions consists of those distributions which do not grow too fast at infinity. Thus, even if we were able to find a solution in the class of all distributions, it would not be physically acceptable because of its bahaviour at infinity.

Non-existence of a good fundamental solution, being a disadvantage from the point of view of calculational efficiency of the theory, might however be thought to be due to too singular character of the δ-type forces. For smooth forces, the theory might still be expected to work and yield smooth solutions.

To see that this is not exactly the case, let us consider an example of forces

$$f_i = -\partial_i \Psi \tag{5.11}$$

where

$$\Psi = \left(\frac{\alpha}{\sqrt{\pi}}\right)^3 e^{-\alpha^2 r^2} \tag{5.12}$$

is a function similar to ϕ but with a different parameter α. These forces are central, with no resultant force or moment of force, and their magnitude as a function of the distance r is

$$f = 2\alpha^2 r e^{-\alpha^2 r^2} \left(\frac{\alpha}{\sqrt{\pi}}\right)^3 . \tag{5.13}$$

The force field (5.11) is infinitely differentiable and, if $\alpha^{-1} \gg a$, describes perfectly smooth distribution of forces on the atomic scale.

Now, applying the Fourier transformation, instead of the eqn (5.7) we obtain:

$$\hat{\Phi}_{ij}(k)\hat{u}_j(k) = -ik_i \hat{\Psi}(k) , \tag{5.14}$$

the solution of which is

$$\hat{u} = -\frac{1}{\lambda+2\mu} \frac{ik_i}{k^2} e^{-k/4\gamma^2} , \tag{5.15}$$

where

$$\frac{1}{\gamma^2} = \frac{1}{\alpha^2} - \frac{1}{\beta^2} . \tag{5.16}$$

If $\alpha<\beta$ - i.e., the forces are diffused over a distance greater than the range of interactions - then $\gamma^2 > 0$, and there exists a smooth solution which, by retransformation of (5.15), is equal to

$$u_i(x) = -\frac{1}{4\pi}\frac{1}{\lambda+2\mu}\partial_i\frac{\operatorname{erf}(\gamma r)}{r} , \tag{5.17}$$

where erf denotes the corresponding error function. If $\alpha = \beta$, there exists a singular solution

$$u_i(x) = -\frac{1}{4\pi}\frac{1}{\gamma+2\mu}\partial_i\frac{1}{r} \tag{5.18}$$

which coincides with the classical solution corresponding to $\Psi(x) = \delta^{(3)}(x)$.

In the case of $\alpha > \beta$, there is no solution in the class of tempered distributions. There exists, in fact, a solution given by the eqn (5.17) with imaginary γ derived from (5.16). It can be checked by direct computation, the integral being, provided that $\alpha^{-1} \neq 0$, very well convergent. This solution, however, grows up exponentially at infinity.

This is not what can be expected on physical grounds. Although a good solution exists when the forces are sufficiently diffused, the necessary degree of diffusion is determined by the range of interactions instead of by the interatomic distance. The range β^{-1} can in principle be made very large so that the inequalities

$$a \ll \alpha^{-1} < \beta \tag{5.19}$$

can be satisfied very well. In spite of that, no acceptable solution exists in this case.

What has been said in this paragraph refers directly to a particular case of an integral equation. Nevetheless, it shows that in formulating nonlocal continuum theories, due attention to the mathematical side of the problem is necessary.

III. PHENOMENOLOGICAL PRINCIPLES OF NONLOCAL ELASTIC CONTINUUM

1. Physical modeling and phenomenological approach

There are two commonly used ways of constructing continuum theory. First of them starts from another model of matter, most frequently the molecular one, and introduces the material continuum as an auxilliary concept. The properties of the continuum are to be derived from the initial model.

The phenomenological approach takes the continuum as an independent model of matter, assuming from the beginning continuum representation of material bodies. The properties of a body are then to be inferred from the general physical laws applied directly to the continuum, combined with some assumptions specifying the phenomenological type of the body.

Both approaches have been historically very fruitful. They both have obvious advantages: in physical modeling one profits from the physical knowledge that one can incorporate into the constructed theory, while the phenomenological approach results in theories showing much simpler conceptual structure with clear interdependence between phenomenological features of the objects and processes under consideration.

In this lecture we shall try to investigate nonlocal continuum theory of elasticity in a more systematic way within the framework of the phenomenological approach. In order to avoid discussing questions connected with the boundary of a nonlocal body, in this chapter we shall assume the medium to be infinite. The very first question we meet here concerns the kind of governing equation one should choose. The almost automatic answer that it is an integral equation is many respects not satisfactory. From the mathematical point of view such an answer tells us almost nothing, unless it is specified in what sense the integrals involved are to be understood. Classical integrals are usually to restrictive, since many singular functions of physical interest cannot be integrated in a classical way. Even if we choose some generalized notions of the integral, we cannot guarantee that a non-differential equation, if acceptable on physical grounds, has to be an integral one or, at least, that can be reasonably written by means of such integrals.

The whole question is not unimportant because the form of governing equation can forejudge physically important features of its solutions. Bearing this in mind, we shall discuss a wide class of linear governing equation which apart from restrictions of direct physical meaning we submit to some mathematical assumptions of rather general character only.

In classical elasticity we have the governing equation of the form

$$L^{class}\,\underline{u} = \underline{f} \tag{1.1}$$

where $\underline{u}$ and $\underline{f}$ denote the displacement and the force fields, respectively, and L^{class} is a differential operator of second order. With index notation, it can be represented as

$$L^{class}_{ij} = \partial_k c_{ijkl} \partial_l \quad , \tag{1.2}$$

where c_{ijkl} denotes the tensor of elastic modulae.

In our phenomenological approach to nonlocal continuum theory we shall investigate equations of the form

$$L\underline{u} = \underline{f} \tag{1.3}$$

or, in index notation,

$$L_{ij} u_j = f_i \tag{1.3a}$$

where L is a certain linear operator. First of all we shall specify the assumptions which make L admissible as a governing operator of (generally nonlocal) continuum theory of elasticity.

2. Basic assumptions on L

A. The displacement field $\underline{u}$ and the force $\underline{f}$ will always be considered tempered (real vector) distributions on the three-dimensional Euclidean space. The operator

$$L \; : \; U \longrightarrow S' \qquad (2.1)$$

is defined on a domain U which need not coincide with the whole space of tempered distributions S' . We assume that U is a linear subset of S' and, as it can depend on L, we do not specify it in advance.

B. We assume the operator L to be continuous in the following sense: whenever a sequence $\underline{u}_1, \underline{u}_2, \underline{u}_3, \ldots \in U$ converges to $\underline{u} \in U$,

$$\underline{u}_1, \underline{u}_2, \underline{u}_3, \ldots \longrightarrow \underline{u}, \qquad (2.2)$$

then the corresponding sequence $\underline{f}_1 = L\underline{u}_1$, $\underline{f}_2 = L\underline{u}_2$, $\underline{f}_3 = L\underline{u}_3$, ... converges to $\underline{f} = L\underline{u}$,

$$\underline{f}_1, \underline{f}_2, \underline{f}_3, \ldots \longrightarrow \underline{f} \quad . \qquad (2.3)$$

The arrows in (2.2) and (2.3) indicate the weak convergence of distributions.

C. The domain U of the operator L contains all functions of the form

$$\underline{a} \sin \underline{k}\underline{x} \quad \text{and} \quad \underline{a} \cos \underline{k}\underline{x} \qquad (2.4)$$

with arbitrary real amplitudes $\underline{a}$ and wave vectors $\underline{k}$.

D. Let $T_{\underline{c}}$ denote the translation operator in S', whose action on a field $\underline{u}$ consists in shifting it by a constant vector $\underline{c}$:

$$T_{\underline{c}}\underline{u}\,(x) = \underline{u}\,(\underline{x} - \underline{c}) \qquad (2.5)$$

To express that the medium under considered action is infinite and homogeneous, we assume that

(i) the domain U is translation-invariant, i.e. whenever $\underline{u} \in U$ then $T_{\underline{c}}\underline{u} \in U$,

(ii) the operator L commutes with the translation operator $T_{\underline{c}}$,

$$LT_{\underline{c}} = T_{\underline{c}}L , \tag{2.6}$$

for arbitrary $\underline{c}$.

Remark. Although we are interested in displacement and force fields which are real, it is convenient to extend the operator to complex-valued distributions. This can be done uniquely by linearity of L. The extended operator L remains real in the sense that whenever applied to a real field it produces a real field. Generally, for any complex $\underline{u}$ from the extended domain of L we have

$$L\operatorname{Re}\underline{u} = \operatorname{Re}L\underline{u} . \tag{2.7}$$

The extended domain, for which we preserve the symbol U, contains all functions of the form

$$\underline{u}(\underline{x}) = \underline{a}\, e^{i\underline{kx}} \tag{2.8}$$

with arbitrary complex amplitudes $\underline{a}$ and arbitrary real wave vectors $\underline{k}$. This follows immediately from the assumption C.

3. Fourier representation of L

Now we make use of the fact that any fields $\underline{u}(\underline{x})$, $\underline{f}(\underline{x})$ which are tempered distributions have well-defined Fourier transforms $\underline{u}(\underline{k})$, $\underline{f}(\underline{k})$ which are also tempered distributions. Let

$$F \;:\; S' \longrightarrow S' \tag{3.1}$$

be the operator representing the Fourier transformation and

$$\hat{U} = F\,[U] \tag{3.2}$$

be the Fourier-image of the domain U. Equation (1.3) can now be expressed as

$$\hat{L}\hat{\underline{u}} = \hat{\underline{f}} \tag{3.3}$$

with the operator

$$\hat{L} \;:\; \hat{U} \longrightarrow S' \tag{3.4}$$

given by the following formula

$$\hat{L} = F^{-1}\,LF \;. \tag{3.5}$$

Remark (3.1). There is one-to-one correspondence between the operators L and $\hat{L}$. Since F and F^{-1} are continuous, the continuity of L is equivalent to that of $\hat{L}$.

Consider now the action of the operator L on a field (2.8). Let $\underline{f}(\underline{x})$ be the corresponding force field. By homogeneity and linearity we have

$$T_{\underline{c}}\underline{f} = T_{\underline{c}}L\underline{u} = LT_{\underline{c}}\underline{u} = Le^{-i\underline{kc}}\underline{u} = e^{-i\underline{kc}}L\underline{u} = e^{-i\underline{kc}}\,\underline{f},$$

with the result that

$$\underline{f}(\underline{x} - \underline{c}) = e^{-i\underline{kc}}\,f(\underline{x}) \tag{3.6}$$

for arbitrary vector $\underline{c}$. From that we conclude

$$\underline{f}(\underline{x}) = \underline{b}\,e^{-i\underline{kx}} \tag{3.7}$$

with another complex amplitude $\underline{b}$ and same wave vector $\underline{k}$.

Since the operator L is linear, there is a linear relation between the complex amplitudes $\underline{a}$ and $\underline{b}$. Such a relation is given by a matrix which, generally, depends on the wave vector k :

$$b_i = \Lambda_{ij}(\underline{k})a_j \quad . \tag{3.8}$$

The matrix $\Lambda_{ij}(\underline{k})$ is uniquely determined by the operator L. From the reality condition (2.7) it follows that

$$\Lambda^*_{ij}(\underline{k}) = \Lambda_{ij}(-\underline{k}) \quad , \tag{3.9}$$

where asterisk denote complex conjugation.

Proposition (3.1). The matrix $\Lambda_{ij}(\underline{k})$ is a continuous function of $\underline{k}$.

Proof: Let $\underline{k}_1, \underline{k}_2, \underline{k}_3, \ldots$ be an arbitrary sequence convergent to a certain wave vector $\underline{k}$,

$$\underline{k}_1, \underline{k}_2, \underline{k}_3, \ldots. \longrightarrow \underline{k} \quad .$$

Then

$$e^{i\underline{k}_1\underline{x}} , e^{i\underline{k}_2\underline{x}}, e^{i\underline{k}_3\underline{x}} , \ldots \longrightarrow e^{i\underline{k}\underline{x}}$$

in the sense of weak convergence of distributions and, by the continuity of L ,

$$\Lambda_{ij}(\underline{k}_1), \quad \Lambda_{ij}(\underline{k}_2) , \quad \Lambda_{ij}(\underline{k}_3), \ldots. \longrightarrow \Lambda_{ij}(\underline{k}) \quad .$$

Proposition (3.2) Let C_o denote the set of continuous functions of compact support, $\hat{C}_o$ - its Fourier image. Then

(i) the operator L can be extended from the set of functions (2.8) onto $\hat{C}_o$ with preserving the continuity,

(ii) the extension of L onto $\hat{C}_o$ is unique,

(iii) the action of $\hat{L}$ on C_o is given by the multiplication

$$\hat{L}_{ij}\, \hat{u}(\underline{k}) = \Lambda_{ij}(\underline{k})\, \hat{u}_j(\underline{k}) \; . \qquad (3.10)$$

Proof: The formula (3.10) defines the continuous extension of $\hat{L}$ and therefore, by virtue of the remark (3.1), the corresponding extension of L. To prove uniqueness of this extension, consider an arbitrary $\hat{\underline{u}} \in C_o$ and the Fourier image of it, $\underline{u} \in \hat{C}_o$. We have

$$\underline{u}(\underline{x}) = \frac{1}{(2\pi)^3} \int d^3k \; e^{i\underline{kx}}\, \hat{\underline{u}}(\underline{k}) \qquad (3.11)$$

in the sense of Riemann integral . A Riemann integral is, by definition, a limit of finite sums. In our case, formula (3.11) express $\underline{u}(\underline{x})$ as a limit of linear combinations of exponential functions (2.8). Hence any u $\in \hat{C}_o$ is a limit of a sequence of such functions. Taking into account the definition of continuity of L we conclude uniqueness of the extension .

Proposition (3.3). If operators L and L′ have the same domain U and the same matrix function $\Lambda_{ij}(\underline{k})$, then L = L′.

Proof: Follows from the fact that any $\hat{\underline{u}}$ is the limit of a sequence of continuous functions of compact support.

4. Energy, stability and the classical limit

The expression for the total deformation energy of a nonlocal elastic medium can be derived from the form of the governing equation. The energy corresponding to diplacements

$\underline{u}(\underline{x})$ produced by force $\underline{f}(\underline{x})$ equals

$$W = \frac{1}{2} \int d^3x \underline{u}\underline{f} = \frac{1}{2} \int d^3x \; \underline{u}L\underline{u} \quad , \tag{4.1}$$

which follows from integrating the elementary work

$$\delta W = \int d^3x \; \delta\underline{u}\underline{f} \quad , \tag{4.2}$$

with making use of linearity of equation (1.1). In Fourier representation, the expresion (4.1) can be written as

$$W = \frac{1}{2} \; \frac{1}{(2\pi)^3} \int d^3k \; u_i \; (\underline{k}) \quad \Lambda_{ij}(\underline{k}) \tag{4.3}$$

E. Now, consider a cyclic deformation process of the form

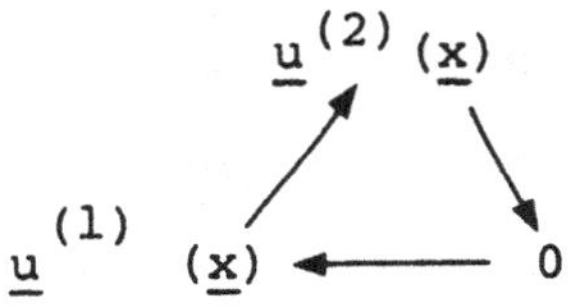

with some displacement fields $\underline{u}^{(1)}(\underline{x})$ and the corresponding force fields $\underline{f}^{(1)}(\underline{x})$, $\underline{f}^{(2)}(\underline{x})$. The medium being elastic, the work done in this process has to be zero:

$$0 = W_{o1} + W_{12} + W_{20} = \frac{1}{2} \int d^3x (\underline{u}^{(2)}\underline{f}^{(1)} - \underline{u}^{(1)}\underline{f}^{(2)}) \; . \tag{4.4}$$

On transforming this relation to Fourier representation and making use of equation (3.10), we obtain:

$$\int d^3k \; u_i^{*(2)}(\underline{k}) \; [\; \Lambda_{ij}(\underline{k}) \; - \; \Lambda_{ji}^{*}(\underline{k})] \; u_j^{(1)}(\underline{k}) \; = \; 0 \tag{4.5}$$

Let us note that the expression (4.3) is well defined for sufficiently many $\underline{u}(\underline{k})$: at least for all the continuous functions of bounded support. Therefore, from the relation (4.5) it follows that

$$\Lambda_{ij}(\underline{k}) = \Lambda^{*}_{ji}(\underline{k}) \quad . \tag{4.6}$$

Taking into account equation (3.9), we have then

$$\Lambda_{ij}(\underline{k}) = \Lambda^{*}_{ij}(-\underline{k}) = \Lambda^{*}_{ji}(\underline{k}) \tag{4.7}$$

F. Moreover, we assume the medium to be stable. As simple argument shows, the stability condition requires the roots $\omega_1^2(\underline{k})$, $\omega_2^2(\underline{k})$, $\omega_3^2(\underline{k})$ of the characteristic equation

$$\det\left(\Lambda_{ij}(k) - \omega^2 \delta_{ij} \right) = 0 \tag{4.8}$$

to be positive for any real $\underline{k} \neq 0$. Thus the matrix $\Lambda_{ij}(\underline{k})$ must be positive definite for $\underline{k} \neq 0$ and, in particular,

$$\det\left(\Lambda_{ij}(\underline{k}) \right) \neq 0 \quad \text{for } \underline{k} \neq 0 \tag{4.9}$$

G. So far we have made no assumptions concerning the relation between nonlocal and classical elasticity. We assume that equation (3.10) agrees with its classical counterpart in the limit $\underline{k} \longrightarrow 0$. Thus

$$\Lambda_{ij}(\underline{k}) = c_{iljm} k_l k_m + o(k^2) \quad \text{when } k \longrightarrow 0 . \tag{4.10}$$

This completes the list of assumptions concerning the operator L.

The last assumption expresses the idea that classical elasticity is to follow from nonlocal theory in the so called long-wave limit.

IV. STRAIN-GRADIENT THEORY OF AN ARBITRARY ORDER

1. The fundamental equation

Now we shall investigate a simple and interesting class of weakly nonlocal elastic media. Instead of L^{class} given by formula III. (1.2), one may try to take more general differential operator in the governing equation. This idea leads to the general fundamental equation of the form

$$P_{ij}(\underline{\partial})u_j = f_j \qquad (1.1)$$

which is a particular case of equation III (1.3a). The symbol $P_{ij}(\underline{\partial})$ represents a tensor-operator which is a polynominal in the partial derivative operators

$$\underline{\partial} = (\partial_1, \partial_2, \partial_3) \quad . \qquad (1.2)$$

This operator will be submitted to the postulates A - G of Chapter III and otherwise arbitrary. For reason which will become clear in section 2 of the present chapter, such a theory is called strain gradient theory of order $\underline{r}$, where $\underline{r}$ is the order of the polynominal $P_{ij}(\underline{\partial})$ in the usual sense, The order $\underline{r}$ will be considered fixed but arbitrary.

The usual tensor notation is not convenient is dealing

with quantities of unspecified order. It can be simplified by the following convention.

Consider an arbitrary tensor quantity of an arbitrary order which is symmetric in a certain group of s indices:

$$a_{\dots i_1 i_2 \dots i_s \dots}$$

Instead of specifying the value of every index in the group, it suffices to state how many indices take the values 1, 2 and 3, respectively. Thus, an arbitrarily large group of symmetric tensor indices can be replaced by three non-negative integers μ_1, μ_2, μ_3. Such a triple will be denoted by a single letter, e.g. $\mu = (\mu_1, \mu_2, \mu_3)$. The quantity

$$|\mu| \overset{[\mathrm{df}]}{=} \mu_1 + \mu_2 + \mu_3 \qquad (1.2)$$

equals the number of tensor indices which correspond to the multi-index μ. For gradient operators of arbitrary orders the above convention allows us to write:

$$\partial^{\mu} \overset{\mathrm{df}}{=} \partial_1^{\mu_1} \partial_2^{\mu_2} \partial_3^{\mu_3} \qquad (1.3)$$

We can consider quantities with arbitrary numbers of multi-indices and/or usual tensor indices.

The operator $P_{ij}(\partial)$ in eqns (1.1) can be conveniently written as:

$$P_{ij}(\partial) = \sum_{|\mu|<r} a_{ij\mu} \partial^{\mu} \qquad (1.4)$$

where the coefficient $a_{ij\mu}$ has two tensor indices i, j and one multi-index μ. For any specified value of $|\mu|$, the quantity $a_{ij\mu}$ is equivalent to a tensor of order $2+|\mu|$ symmetric in the last $|\mu|$ indices.

2. Conditions for the coefficients $a_{ij\mu}$

For a homogeneous medium the coefficients $a_{ij\mu}$ do not depend on $\underline{x}$. The matrix $\Lambda_{ij}(\underline{k})$ associated with the operator $P_{ij}(\underline{\partial})$ according to the definition given in section III. 3 can be expressed as

$$\Lambda_{ij}(k) = \sum_{|\mu|<r} a_{ij\mu} (i\underline{k}) \quad . \tag{2.1}$$

All the coefficients $a_{ij\mu}$ should evidently be real, so that

$$\Lambda_{ij}(\underline{k}) = \sum_{|\mu|<r} a_{ij\mu} (-i\underline{k}) \tag{2.2}$$

From the relation III.(4.7) one obtains

$$a_{ij\mu} = (-1)^{\mu} a_{ji\mu} \tag{2.3}$$

In consequence, if the order $|\mu|$ of the corresponding term is even, then the coefficients $a_{ij\mu}$ are symmetric in i, j and antisymmetric on the contrary.

Apart from relations (2.3), for a given anisotropy class the coefficients $a_{ij\mu}$ have to satisfy the relations required by the corresponding point group. In particular, if the medium is centrosymmetric, all the coefficients corresponding to odd order $|\mu|$ vanish.

Moreover, the coefficients $a_{ij\mu}$ have to be chosen in such a way, that the matrix (2.1) be, according to the postulate of stability, positive definite for all $k \neq 0$. The postulate III.F, concerning the correspondence with classical elasticity, eliminates the terms of order 0 and 1:

$$a_{ij\mu} = 0 \text{ for } |\mu| < 1 \ . \tag{2.4}$$

The last condition completes the list of restrictions that have to be imposed on the coefficients $a_{ij\mu}$.

3. Energy

The general expression III. (4.3) for global energy of the medium takes now the form

$$W = \frac{1}{2}\frac{1}{(2\pi)^3}\int d_3k\ \hat{u}_i\ (\underline{k})\quad P_{ij}\ (\underline{k})\hat{u}_j(\underline{k}) \tag{3.1}$$

In consequence, the global energy of an infinite medium for a given displacement field is uniquely determined by the form of the operator $P_{ij}(\partial)$. It does not, however, allow us to determine the energy density uniquely. In fact, introducing an energy density ω so that

$$W = \int d_3 x\omega \tag{3.2}$$

we obtain an expression equivalent to (2.3), provided that the equation for δW resulting from (3.2) agress with eqn (2.2) . The necessary and sufficient condition is (3.3)

$$P_{ij}(\partial)u_j = \frac{\partial W}{\partial u_j} \quad , \tag{3.3}$$

where the last symbol denotes the functional derivative

$$\frac{\delta\omega}{\delta u_i} = \sum_\mu (-)^\mu \partial^\mu \frac{\partial\omega}{\partial u_{i,\mu}} . \tag{3.4}$$

In this case, the energy density ω differs from $\frac{1}{2} u \cdot f$ by a divergence-type terms which does not affect the global energy.

The energy density ω can be subjected to further requirements, such as invariance with respect to rigid translations and rotations of medium or, for homogeneous deformations, correspondence to the classical theory of elasticity. With no initial stress present, the invariance requirement eliminates dependence ω on u_i and ω_{ij} so that the energy density depends only on ε_{ij} and the derivatives of u_i of second and possibly, higher orders:

$$\omega = \omega(\varepsilon_{ij}, u_{i,\mu}) , \quad |\mu| \geqslant 2 \tag{3.5}$$

the symbols ω_{ij} and ε_{ij} have the standard meaning:

$$\omega_{ij} = \frac{1}{2} (u_{i,j} - u_{j,i}) , \tag{3.6}$$

$$\varepsilon_{ij} = \frac{1}{2} (u_{i,j} + u_{j,i}) . \tag{3.7}$$

The gradients of any non-zero order of the rotation tensor ω_{ij} can be entirely expressed by the gradients of the corresponding order of the strain tensor ε_{ij}. In fact, we can easily check that the equation

$$\omega_{ij,k} = \varepsilon_{ik,j} - \varepsilon_{jk,i} \tag{3.8}$$

holds identically. Making use of this equation, we can write:

$$u_{i,kj} = \varepsilon_{ij,k} + \omega_{ij,k} = \varepsilon_{ij,k} + \varepsilon_{ik,j} - \varepsilon_{jk,i} \tag{3.9}$$

which enables us to express the second order gradients of the displacement vector u_i by the first gradient of the strain tensor ε_{ij}. By differentiating eqn (3.9) the analogous equations for higher order gradients can be obtained. This leads us to the conclusion that energy density (3.5) can be expressed as a function of the strain tensor and its gradients of different orders:

$$\omega = \omega(\varepsilon_{ij}, \varepsilon_{ij,\mu}), \quad |\mu| > 1 . \tag{3.10}$$

Given an expression for the energy density, the form of the operator $P_{ij}(\partial)$ is determined uniquely by eqn (3.3). Therefore, in phenomenological formulation of the strain gradient theory, it is often convenient to begin with consideration of the energy density.

The energy of a finite body cannot be unambiguously determined without further investigation. This is bacause the surface energy may contain terms non-equivalent to any volume integral. For the present purposes, knowledge of the energy of an unbounded medium will suffice.

4. The operator $P_{ij}(\partial)$ and length parameters for and isotropic medium

Now we shall investigate the question as to what effect the higher order terms have on solutions to eqns (1.1). We shall begin with an isotropic medium.

In that case, the most general form of the operator $P_{ij}(\partial)$ is

$$P_{ij}(\partial) = - a(\Delta) \; (\Delta\delta_{ij} - \partial_i\partial_j) - b(\Delta)\partial_i\partial_j \; , \qquad (4.1)$$

where $a(\Delta)$ and $b(\Delta)$ are polynomials in the Laplace operator Δ . Equation (1.1) takes the form

$$- a(\Delta) \; \Delta u - [b(\Delta) - a(\Delta)] \text{ grad div } u = f \; . \qquad (4.2)$$

From correspondence with the classical theory, we have

$$a(0) = \mu \; , \qquad b(0) = \lambda + 2\mu \; , \qquad (4.3)$$

where λ and μ represent Lamé constants.

Let p and q be the orders of the polynomials $a(\Delta)$ and $b(\Delta)$ respectively, and

$$\beta_2^2 \, , \beta_2^2 \, , \; \dots \; , \beta_p^2 \, ,$$

$$\beta_1'^2 , \; \beta_2'^2 , \; \dots \; , \; \beta_q'^2 \qquad (4.4)$$

their roots, each taken according to its multiplicity. Then we can write:

$$a(\Delta) = \text{const.} \; (\beta_1^2 - \Delta) \; (\beta_2^2 - \Delta) \; \dots \; (\beta_p^2 - \Delta) \; ,$$

$$b(\Delta) = \text{const.} \; (\beta_1'^2 - \Delta) \; (\beta_2'^2 - \Delta) \; \dots \; (\beta_q'^2 - \Delta) . \qquad (4.5)$$

The roots (4.4) are, in general, complex. They obey, however, the following restrictions:

(a) none of them is equal to zero;

(b) none of $\beta_r'^2$ or β_r^2 is a real negative number;

(c) non-real roots of $a(\Delta)$, as well as those of $b(\Delta)$, occur in mutually conjugate pairs.

The restriction (a) follows from correspondence with classical theory of elasticity. In fact, from (4.5) we have

$$a(0) = \text{const.}\ \beta_1^2\ \beta_2^2 \dots \beta_p^2\ ,$$

$$b(0) = \text{const.}\ \beta_1'^2\ \beta_1'^2 \dots\ \beta_q'^2\ . \qquad (4.6)$$

If one or more of the roots were zero, then according to (4.3), one or both Lamé constants would be zero; this contradicts the classical theory.

The restriction (b) is a consequence of the stability condition. According to eqn (4.1), the matrix (3.7) for an isotropic medium can be written as

$$P_{ij}(ik) = a(-k^2)\ (k^2\ \delta_{ij} - k_i k_j) + b(-k^2) k_i k_j\ . \qquad (4.7)$$

Thus, this matrix is positive-definite only when

$$a(-k^2) > 0 \qquad \text{and} \qquad b(-k^2) > 0\ . \qquad (4.8)$$

If a real negative root of any of the two polynomials existed, it would contradict the inequalities (4.8) for certain real wave vectors k.

The conditions (a) and (b) imply that β's themselves can be so chosen that

$$\mathrm{Re}\,\beta_r > 0\,, \qquad \mathrm{Re}\,\beta'_s > 0 \tag{4.9}$$

for any r and s.

The condition (c) is an immediate consequence of the fact that the coefficients of the polynomials $a(\Delta)$ and $b(\Delta)$ are real.

All the β's have the dimension of inverse length and therefore can be used as a convenient set of length parameters. The number of independent real length parameters equals the total number of the roots β_i and β'_j : a real root determines one, and a pair of mutually conjugate complex roots determine two such parameters.

The classical elastic constants and the roots (4.4) determine completely the polynomials $a(\Delta)$ and $b(\Delta)$, and, consequently, the detailed form of eqn (4.2).

In fact, according to eqns (4.5) and (4.3), we have the following representations

$$a(\Delta) = \frac{\mu}{\beta_1^2\beta_2^2\ \ldots\ \beta_p^2}(\beta_1^2 - \Delta)(\beta_2^2 - \Delta)\ldots(\beta_p^2 - \Delta)$$

$$= \mu(1-\beta_1^{-2}\Delta)(1-\beta_2^{-2}\Delta)\ldots(1-\beta_p^{-2}\Delta), \tag{4.10}$$

$$b(\Delta) = \frac{\lambda + 2\mu}{\beta_1'^2\beta_2'^2\ \ldots\ \beta_q'^2}(\beta_1'^2 - \Delta)(\beta_2'^2 - \Delta)\ldots(\beta_q'^2 - \Delta)$$

$$= (\lambda + 2\mu)(1-\beta_1'^{-2}\Delta)(1-\beta_2'^{-2}\Delta)\dots(1-\beta_q'^{-2}\Delta) .$$

5. The general form of solutions to homogeneous equations for isotropic media

Let us consider eqn (4.2) in a region where $f = o$:

$$a(\Delta)\Delta u + (b(\Delta) - \alpha(\Delta))\text{grad div } u = 0 \tag{5.1}$$

The displacement field can be decomposed into two parts

$$u = v + w \tag{5.2}$$

so that

$$\text{div } v = 0 \text{ and } \text{curl } w = 0 . \tag{5.3}$$

The fields v and w have to satisfy the eqns

$$a(\Delta)\Delta^2 v = 0 \quad \text{and} \quad b(\Delta)\Delta^2 w = 0 \tag{5.4}$$

In the classical theory of elasticity, where $a(\Delta)$ and $b(\Delta)$ are simply constants, the fields v and w (and u, in consequence) have to be biharmonic:

$$\Delta^2 v^{class} = 0 \quad \text{and} \quad \Delta^2 w^{class} = 0 \tag{5.5}$$

Thus, according to eqns (5.4), any classical solution is acceptable as a particular solution to the equations of the strain gradient theory[x)]. The non-classical solutions in

---

x) If v and w are biharmonic, then $a(\Delta)\Delta v = a\Delta v$, $b(\Delta)\Delta w = b\Delta w$.
As with the substitution of (5.2) eqn (5.1) becomes
$$a(\Delta)\Delta v + b(\Delta)\Delta w = 0 ,$$
it is satisfied if the classical equation

which we are interested can be found by considering the equation

$$\Delta u = \beta^2 u \ , \tag{5.6}$$

where β^2 is a constant. Taking into account eqns (5.3) and (5.4), we see that if either

$$a(\beta^2) = 0 \quad \text{and} \quad \operatorname{div} u = 0 \ , \tag{5.7}$$

or

$$b(\beta^2) = 0 \text{ and } \operatorname{curl} u = 0 \ , \tag{5.8}$$

then the solution u to eqn (5.6) satisfies eqn (5.1). In that case, the constant β^2 in eqn (5.6) must be equal to one of the roots (4.4). We may conclude that a sum of the form

$$u = u^{class} + \sum_s u^{(s)} , \tag{5.9}$$

where u^{class} is a classical solution and the $u^{(s)}$s satisfy either

$$\Delta u^{(s)} = \beta_s^2 u^{(s)} , \ \operatorname{div} u^{(s)} = 0, \tag{5.10}$$

or

$$\Delta u^{(s)} = \beta_s'^2 \ u^{(s)} , \ \operatorname{curl} u^{(s)} = 0 \ , \tag{5.11}$$

is always a solution to eqn (5.1).

---

x)

$$a \Delta v + b \Delta w = 0$$

is satisfied. Hence, in the isotropic case, classical solutions are always solutions to strain gradient theory.

In the case in which the polynomials $a(\Delta)$ and $b(\Delta)$ have no multiple roots, the inverse statement is also true: any solution to eqn (5.1) is of the form (5.9). This can be proved by induction with respect to the number of factors in the representations (4.10). Thus, in that case the formula (5.9) represents the most general form of solution to the equations of the strain gradient theory.

If any of polynomials $a(\Delta)$ and $b(\Delta)$ has a multiple root, the solution given by the formula (5.9) is not of the most general form. In this case, for any multiple root we have a family of particular solutions to eqns (5.1)

$$u, \frac{\partial}{\partial\beta}u, \ldots, \frac{\partial^{m-1}}{\partial\beta^{m-1}}u, \tag{5.12}$$

where u satisfies eqns (5.10) or (5.11), and m represents the multiplicity of the root β. This follows from the observation that if u is a solution to equation (5.6), then

$$(\beta^2 - \Delta)^{l+1} \frac{\partial^l}{\partial\beta^l}u = 0 \tag{5.13}$$

for an arbitrary l. The particular solutions of the form (5.12) for all multiple roots should be taken into account.

6. The three-dimensional fundamental solution for an isotropic medium

We consider first the fundamental solution $G_{ij}(x)$ defined as the solution of the equation:

$$-a(\Delta)G_{ij,kk} - (b(\Delta)-a(\Delta))G_{kj,ki} = \delta^{(3)}(x)\delta_{ij} \tag{6.1}$$

submitted to the condition of vanishing at infinity. The symbols $\delta^{(3)}(x)$ and δ_{ij} denote the three-dimensional Dirac delta and Krönecker delta, respectively. This solution can easily be found by making use of the Fourier transformation. It can be represented by the following Fourier integral

$$G_{ij}(x) = \frac{1}{(2\pi)^3} \int_{-\infty}^{\infty} d_3k \left[\frac{k^2\delta_{ij}-k_ik_j}{a(-k^2)} + \frac{k_ik_j}{b(-k^2)} \right] \frac{e^{ikx}}{k^4}, \tag{6.2}$$

where k^{-4} is to be understood in the following sense

$$\frac{1}{k^4} = \frac{1}{2}\left[\frac{1}{(k+i0)^4} + \frac{1}{(k-i0)^4}\right]. \tag{6.3}$$

After performing angular integrations, we obtain the expression

$$G_{ij}(x) = \frac{1}{(2\pi)^2 i} \int_{-\infty}^{\infty} dk \left[\frac{k^2\delta_{ij}+\partial_i\partial_j}{a(-k^2)} - \frac{\partial_i\partial_j}{b(-k^2)} \right] \frac{e^{ikr}}{k^3 r}. \tag{6.4}$$

This integral can be computed conveniently by passing to the complex k-plane. Because $r > 0$, the integration contour can be closed in the upper half-plane. Then the contributions to the integral (6.2) arise from:

1) the pole $k=0$; according to the formula (6.3) this contribution is given by only a half of the residuum;

2) the poles at the roots of the polynomials $a(-k^2)$ and $b(-k^2)$ with $\mathrm{Im}k > 0$. These poles are at the points

$$k = i\beta_r\ , \qquad r = 1, 2, \ldots, p, \tag{6.5}$$

and

$$k = i\beta'_s\ , \qquad s = 1, 2, \ldots, q \tag{6.6}$$

with β'_s restricted by inequalities (4.9).

The above procedure is equivalent to evaluating the integral (6.4) as

$$\frac{1}{2}\int_{C_0} dk \ldots + \int_{C_1} dk \ldots , \tag{6.7}$$

where the contours C_o and C_1 are chosen as in Fig.1. The contour C_1 encloses all the

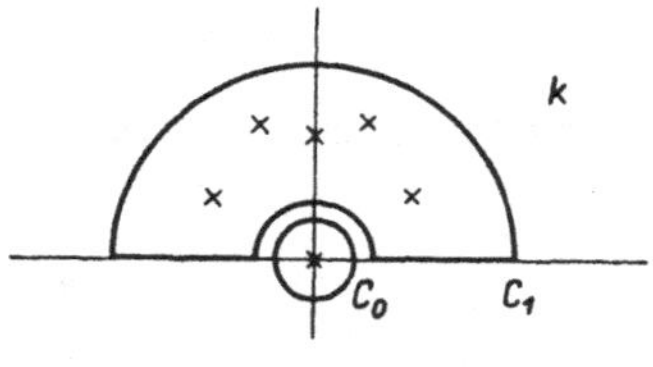

FIG. 1.

poles (6.5) and (6.6) and the contour C_o encircles only the pole at $k = 0$. Summing up the above-mentioned contributions, we obtain

$$G_{ij} = \frac{1}{4\mu}\frac{1}{\mu}\delta_{ij}\left(\frac{1}{r} - \sum_s \alpha_s \frac{e^{-\beta_s r}}{r}\right) - \frac{1}{4\mu}\frac{1}{\mu}\partial_i\partial_j\left(\frac{r}{2} + \sum_s \frac{1}{\beta_s^2}\left(\frac{1}{r} - \alpha_s \frac{e^{-\beta_s r}}{r}\right)\right) +$$

$$+ \frac{1}{4\pi}\frac{1}{\lambda+2\mu}\partial_i\partial_j\left(\frac{r}{2} + \sum_s \frac{1}{\beta_s'^2}\left(\frac{1}{r} - \alpha'_s \frac{e^{-\beta'_s r}}{r}\right)\right) , \tag{6.8}$$

where the symbols α_s and α'_s are defined as

$$\alpha_s = \prod_{r \neq s} \frac{\beta_r^2}{\beta_r^2 - \beta_s^2}, \quad \alpha'_s = \prod_{r \neq s} \frac{\beta_r'^2}{\beta_r'^2 - \beta_s'^2}. \tag{6.9}$$

By making use of the identities

$$\partial_i \partial_j r = \frac{\delta_{ij}}{r} - \frac{x_i x_j}{r^3},$$

$$\partial_i \partial_j \frac{1}{r} = -\frac{4\pi}{3}\delta^{(3)}(x)\delta_{ij} - \frac{\delta_{ij}}{r^3} + \frac{3x_i x_j}{r^5}, \tag{6.10}$$

$$\partial_i \partial_j e^{-\beta r}\frac{1}{r} = -\frac{4\pi}{3}\delta^{(3)}(x)\delta_{ij} - \frac{e^{-\beta r}}{r}\left(\frac{1}{r} + \frac{\beta}{r}\right)\delta_{ij} + \frac{e^{-\beta r}}{r}\left(\frac{3}{r^2} + \frac{3\beta}{r} + \beta^2\right) \cdot \frac{x_i x_j}{r}$$

the angular dependence of G_{ij} can be demonstrated. From the point of view of bahaviour as $r \to \infty$, three types of terms in the fundamental solution (6.8) can be distinguished

1) the terms of order $1/r$; these terms form the classical fundamental solution:

$$G_{ij}^{class} = \frac{1}{4\pi}\frac{1}{\mu}\frac{\delta_{ij}}{r} - \frac{1}{4\pi}\left(\frac{1}{\mu} - \frac{1}{\lambda+2\mu}\right)\partial_i \partial_j \frac{r}{2}$$

$$= \frac{1}{8\pi r}\frac{1}{\mu}\left(\delta_{ij} + \frac{x_i x_j}{r^2}\right) + \frac{1}{\lambda+2\mu}\left(\delta_{ij} - \frac{x_i x_j}{2}\right); \tag{6.11}$$

2) the terms of order $1/r^3$

$$- \frac{1}{4\pi}\left(\frac{1}{\mu} \sum_s \frac{1}{\beta_s^2} - \frac{1}{\lambda+2\mu} \sum_s \frac{1}{\beta_s'^2}\right) \partial_i \partial_j \qquad (6.12)$$

$$= \frac{1}{4\pi} \left(\frac{1}{\mu} \sum_s \frac{1}{\beta_s^2} - \frac{1}{\lambda+2\mu} \sum_s \frac{1}{\beta'^2}\right) \left(\frac{\delta_{ij}}{r^3} - \frac{3x_i x_j}{r^5}\right) ;$$

3) the exponential terms which, by enqualities (4.9) decrease at infinity.

Hence, We can write

$$G_{ij} = G_{ij}^{class} + \frac{1}{4\pi r^2}\left(\frac{1}{\mu} \sum_s \frac{1}{\beta_s^2} + \frac{1}{\lambda+2\mu} \sum_s \frac{1}{\beta_s'^2}\right) \left(\delta_{ij} - \frac{3x_i x_j}{r^5}\right) \qquad (6.13)$$

+ a finite number of exponentially decreasing terms.

For $r \to 0$, the classical fundamental solution exhibits a singularity of order $1/r$. This no longer holds in the case of the strain gradient theory. Taking into account the identities

$$\sum \frac{1}{\beta_s^2} (1-\alpha_s) = 0, \quad \sum \frac{1}{\beta_s'^2} (1-\alpha_s') = 0 \qquad (6.14)$$

and

$$\sum \alpha_s = 1 , \quad \sum \alpha_s' = 1 , \qquad (6.15)$$

it can be seen from the eqn (6.8) that the terms of order $1/r$ cancel each other. Thus, the fundamental solution in the strain gradient theory is continuous at $r = 0$.

The actual form of the singularity at $r=0$ depends on

p and q, the orders of polynomials $a(\Delta)$ and $b(\Delta)$. If $p > 1$ and $q > 1$, the gradients of G_{ij} up to second order, $\partial_k G_{ij}$ and $\partial_l \partial_k G_{ij}$, are continuous. If $p > 2$ and $q > 2$, then the gradients up to fourth order are continuous, and so on. In fact, if $p > 1$ and $q > 1$, the following equations are identically valid:

$$\sum_s \beta_s^{2m} \alpha_s = 0 \quad \text{for} \quad m = 1, 2, \dots, p-1,$$
$$\sum_s \beta_s'^{2m} \alpha_s' = 0 \quad \text{for} \quad m = 1, 2, \dots, q-1, \tag{6.16}$$

so that sufficiently low odd powers of r in the series representing the solution (6.8) are cancelled. Specifically, the following asymptotic expression is valid for small r:

$$G_{ij} = \frac{1}{4\pi}\frac{1}{\mu}(\Delta\delta_{ik} - \partial_i\partial_j)\sum_s\left(\frac{\alpha_s}{\beta_s^2}\frac{\mathrm{sh}\beta_s r}{r} - \frac{\beta_s^{2p}\alpha_s}{(2p+2)} r^{2p+1}\right)$$
$$- \frac{1}{4\pi}\frac{1}{\lambda+2\mu}\partial_i\partial_j\sum_s\left(\frac{\alpha_s'}{\beta_s'^2}\frac{\mathrm{sh}\beta_s' r}{r} - \frac{\beta_s'^{2q}\alpha_s'}{(2q+2)!} r^{2q+1}\right) + \tag{6.17}$$
$$+ O(r^{2p+1}) + O(r^{2q+1})$$

It follows from the above expression that up to order

$$2\min(p,q) - 2 \tag{6.18}$$

the gradients of the fundamental solution G_{ij} are continuous.

V. NONLOCAL MODELS OF DISCRETE STRUCTURES. PSEUDOCONTINUUM

1. Idea

From the atomistic point of view the set of functions admitted in continuum theory is definitely too large. For example, the displacement field u(x) can be interpreted only at those points at which the atoms are located, i.e. on a discrete set of points

$$x = X_n \quad .$$

Generally, there is a lot of functions of the continuous argument x which take some prescribed values $u(X_n) = u_n$ at a discrete set $x = X_n$. In particular, there are many functions u(x) which are not identically zero but vanish at any $x = X_n$. Such functions cannot be reasonably interpreted in atomistic terms. Moreover, the physical meaning of rapidly oscilating functions of position is always doubtful.

This is a fundamental difficulty of any continuum theory. It suggests the idea of restricting the set of admissible functions from the very beginning and to carry out all the mathematical considerations for this restricted set of functions. This set should contain only the functions which are necessary to describe all possible displacements of the atoms. The corresponding model of material medium differs considerably from the continuous one: contrary to the latter case the number of degrees of freedom per unit volume is finite.

The problem of constructing the required set of functions is essentially an interpolation problem. The set of interpolating functions will be denoted by PC. Without further assumptions this set cannot be determined uniquely. We shall discuss three methods of constructing the set PC. Under some additional conditions, they turn out to be equivalent to each other.

For the sake of lucidity, we shall first discuss these methods for scalar fields on a linear chain of atoms. Afterwards we apply this procedure to vector fields on a three-dimensional primitive lattice. The state of a non-primitive structure can be described by multiplets of fields defined on the corresponding Bravais lattice, e.g. one field for each atom in the elementary cell.

2. The sampling function method

Let $\underline{a}$ denote the distance netween neighbouring atoms of a linear chain and let f be a certain quantity taking some values f_n on atoms located at $x = na$ (n - an arbitrary integer).

Let $S(x)$ be a function of the continuous variable x such that

$$S(na) = \begin{cases} 1 \text{ for } n = 0, \\ 0 \text{ for } n \neq 0. \end{cases} \qquad (2.1)$$

Any function satisfying (2.1) will be called a sampling function.

Provided that a sampling function has been chosen, one can uniquely associate a function $f(x)$ of the continuous variable x to a function f_n of the discrete variable n:

$$f(x) = \sum_n S(x - na) f_n \ . \qquad (2.2)$$

This formula makes sense whenever the series involved is convergent. Because of (2.1) the function $f(x)$ has the interpolation property

$$f(na) = f_n \ . \qquad (2.3)$$

The function

$$S(x) = \frac{a}{\pi x} \sin \frac{\pi x}{a} \qquad (2.4)$$

provides an example of a smooth (analytic) sampling function.

3. The Fourier transformation method

Under some restrictions of its behaviour at infinity,

the discrete function f_n can be represented as a Fourier integral

$$f_n = \frac{1}{2\pi}\int_{-\pi/a}^{\pi/a} dk e^{ikan}\hat{f}(k) \tag{3.1}$$

with an appropriate function (or, more generally, distribution) $\hat{f}(k)$. The function $\hat{f}(k)$ is uniquely determined by the function f_n:

$$\hat{f}(k) \doteq \frac{1}{a}\sum_{n=-\infty}^{\infty} e^{-ikan} f_n \ . \tag{3.2}$$

On the other hand, by writing x instead of an in (3.1), one can define a function $f(x)$ of the continuous variable x:

$$f(x) = \frac{1}{2\pi}\int_{-\pi/a}^{\pi/a} dk\, e^{ikx}\hat{f}(k) \ . \tag{3.3}$$

The function $f(x)$ has the interpolation property (2.3)

The Fourier transformation method is equivalent to the sampling function method provided that the sampling function (2.4) is chosen . It follows from the fact that to the function $S(x)$ given by the eqn (2.4) there corresponds the function $\hat{S}(k) = 1$, i.e.,

$$S(x) = \frac{1}{2\pi}\int_{-\pi/a}^{\pi/a} dk\, e^{ikx} \ . \tag{3.4}$$

4. The direct interpolation method

According to the Paley-Wiener-Schwartz theorem, any function of the form (3.3), with an arbitrary distribution $\hat{f}(k)$, can be continued to the complex x-plane as an entire analytic function. Moreover, it satisfies the inequality

$$|f(x)| < C(1 + |x|)^N e^{\pi/a \operatorname{Im} x} \qquad (4.1)$$

for some constants C and N. This inequality is also sufficient for an entire analytic function to have a representation of the form (3.3).

In terms of the theory of analytic functions, f(x) is an entire analytic function of order not greater than 1 and type π/a. This class of functions can be identified with PC. Then the problem of fitting a function $f \in PC$ to given values f_n can be stated directly as an interpolation problem. If $f_n \equiv 0$, then

$$\int_{-\pi/a}^{\pi/a} dk\, e^{ikan} \hat{f}(k) = 0 \qquad (4.2)$$

and $\hat{f}(k)$ must be of the form

$$\hat{f}(k) = P\left(\frac{d}{dk}\right) [\delta(k-\pi/a) - \delta(k+\pi/a)] , \qquad (4.3)$$

where P(.) is a polynomial. Therefore any function $f \in PC$ such that $f_n \equiv 0$ must be of the form

$$f(x) = W(x) \sin\frac{\pi x}{a} , \qquad (4.4)$$

where W(x) is a polynomial. Thus for the functions f(x) and f_n which tend to zero at infinity, the solution of the interpolation problem is unique.

5. The three-dimensional generalization

The above considerations can be generalized directly to multi-dimensional lattices. The atoms of a primitive lattice are labelled by integer multiplets n, and their positions are given by

$$X(n) = An , \qquad (5.1)$$

where A is a certain matrix. The sampling function S(x) should have the property

$$S(An) = \begin{cases} 1 & \text{for} \quad n = 0, \\ 0 & \text{for} \quad n \neq 0. \end{cases} \qquad (5.2)$$

Instead of integrating over the interval $[\pi/a, -\pi/a]$ in (3.1) and (3.3), one should integrate over the first Brillouin zone, e.g.

$$f_n = \frac{1}{(2\pi)^3} \int_{BZ} d^3k e^{ikx} \hat{f}(k) \qquad (5.3)$$

and

$$f(x) = \frac{1}{(2\pi)^3} \int_{BZ} d^3k e^{ikx} \hat{f}(k) . \qquad (5.4)$$

Although it is possible to choose some other equivalent domains of integration, the choise of the BZ has the advantage of retaining the point symmetry of the corresponding Bravais lattice.

The sampling function corresponding to the Fourier transformation method is given as

$$S(x) = \frac{1}{(2\pi)^3} \int_{BZ} d^3k e^{ikx} . \qquad (5.5)$$

The formulae (5.3), (5.4) and (2.2) can be directly adapted for multi-component quantities, e.g. for displacement fields, force fields etc..

6. Some properties of PC-functions

The functions PC have been defined as the entire analytic functions satisfying inequality (4.1). In particular, all polynomials are PC.

The following simple properties can be checked directly:

- if $u \in PC$, then all the derivatives $\partial^\mu u \in PC$,
- if $u \in PC$ and the convolution $u*v$ exists then $u*v \in PC$ (but it can happen that $u \cdot v \notin PC$).

The following equations hold:

$$\Omega \sum_n f_n = \int d^3x f(x) , \qquad (6.1)$$

$$\Omega \sum_n f_n g_n = \int d^3x f(x) g(x), \qquad (6.2)$$

$$\sum_{n,n'} u_n \Phi^{nn'} w^{n'} = \int d^3x d^3x' u(x) \Phi(x,x') w(x'), \qquad (6.3)$$

where Ω denotes the volume of the primitive cell.

In particular, it follows from (6.2) that the kinetic energy equals

$$T = \frac{m}{2} \sum_n v_n^2 = \frac{\rho}{2} \int d^3x v^2(x) , \qquad (6.4)$$

where $\rho = m/\Omega$. On the other hand, the potential energy of a crystal can be expressed as

$$U = \frac{1}{2}\sum_{n,n'} u_{in}\Phi_{ij}^{nn'}u_{jn'} = \frac{1}{2\Omega^2}\int d^3x d^3x' u_i(x)\Phi_{ij}(x,x')u_j(x') . \tag{6.5}$$

The last expression is formally similar to the corresponding expression of non-local (integral) theory. The corresponding expression in (6.4) is also similar to the continuum one. The expression (6.4) and (6.5) for the kinetic and potential energies allow us to develop the theory in continuum-like form. The equations obtained, however, describe the crystal exactly.

In the same way as in continuum theory, these equations can be generalized to include dislocation fields. We shall omit here the details of this procedure.

7. Equivalence of operators in PC

Consider two linear operators L and L′. If for any $u \in PC$

$$Lu = L'u \tag{7.1}$$

these operators will be called equivalent. In fact, from the point of view of PC they are identical, and in any calculation one can put L′ instead of L. This circumstances gives the PC-theory great flexibility.

As an example, consider a differential operator

$$Lu = P(\partial)u \ , \tag{7.2}$$

where $P(\partial)$ is a certain polynomial. Let

$$\Psi(x) = \frac{1}{(2\pi)^3} \int_{BZ} d^3k P(ik) e^{ikx} \ ; \tag{7.3}$$

then we define another operator

$$L'u = \Psi * u \ , \tag{7.4}$$

which is an integral operator. Moreover, let

$$\tilde{\Psi}(x,n) = \Omega\Psi(x-An) \tag{7.5}$$

and

$$L''u = \sum_n \tilde{\Psi}(x,n) u(An) \ . \tag{7.6}$$

All the operators L, L', L'' are equivalent to each other. This shows that in PC there are equivalencies between differential, integral and discrete operators.

8. Pseudocontinuum vs. discrete lattice. Differential models

The pseudocontinuum approach, when applied to crystal lattice with given force constants, produces the same physical result as conventional discrete lattice (DL) theory. This, in particular, follows from the formulae (6.4) and (6.5), if we take into account the one-to-one correspondence between force constants and PC-kernels,

$$\Phi_{ij}^{n,n'} \longleftrightarrow \Phi_{ij}(x,x') \ , \tag{8.1}$$

which holds precisely under physically plausible conditions of vanishing at infinity. In this sense, there is an isomorphism between PC and DL. The equations of motion of a discrete lattice

$$m\ddot{u}_i^n - \sum_{n'} \Phi_{ij}^{nn'} u_j^{n'} = f_i^n \tag{8.2}$$

and the corresponding pseudocontinuum,

$$\rho\ddot{u}_i(x) - \int d_3x' \Phi_{ij}(x,x') u_j(x') = f_i(x) \tag{8.3}$$

contain the same physical information.

The mathematics of PC, however, is much more flexible than that of DL. In the framework of PC one can use, apart from discrete operators, also other equivalent forms of operators without loosing the precise meaning of them. Therefore, the calculational efficiency of PC can be much higher than that of DL.

Another, and perhaps even more important circumstance consists in the following. The lattice theory does not provide us with specific set of force constants. These either have to be determined from other physical considerations, or treated as (infinitely many) phenomenological parameters. In any case, establishing a tractable set of force constants involves some approximations.

One of the most popular simplifications in DL is a finite (and, as a matter of fact, very short) range of interactions. It is equivalent to approximating the exact dispersion curves by a combination of trigonometric functions.

On the other hand, it is possible to achieve substantial simplification is a different manner, namely by choosing the force constants of the lattice so that the integral operator in (8.3) is equivalent to a differential one. Then, instead of (8.3) we have a differential equation of the form

$$\rho\ddot{u}_i(x) - P_{ij}(\partial)u_j(x) = f_i(x) \ . \qquad (8.4)$$

This PC-equation describes exactly the dynamics of a certain crystal lattice. The force constants of this lattice do not vanish identically for great distances: instead of that, they tend to zero with some negative power of the distance. In the k-space, this corresponds to approximating the dispersion curves by a polynomial instead of trigonometric functions, the polynomial being the simpler but not necessarily the worse approximation.

VI. NONLOCAL OPERATORS AND FUNDAMENTAL SOLUTIONS. SINGULAR ORDER

1. The concept of singular order

Let us consider an infinite homogeneous nonlocal elastic medium. The governing operator for this medium will be denoted by L and assumed to satisfy the postulates A - F given in Chapter III.

Let p be a real number. We define the following quantity

$$||L||_p = \frac{1}{(2\pi)^3} \int d^3k(1+k^2)^{-p/2} \mathrm{tr}\Lambda(\underline{k}), \tag{1.1}$$

where

$$\mathrm{tr}\Lambda(\underline{k}) = \Lambda_{11}(\underline{k}) + \Lambda_{22}(\underline{k}) + \Lambda_{33}(\underline{k}) =$$
$$= \omega_1^2(\underline{k}) + \omega_2^2(\underline{k}) + \omega_3^2(\underline{k}) > 0 \tag{1.2}$$

From the definition (1.1) it follows immediately that

$$||L||_{p'} < ||L||_p \quad \text{for} \quad p' > p \tag{1.3}$$

Therefore, the set of numbers p for which the inequality

$$||L||_p < \infty \tag{1.4}$$

holds, can conveniently be characterized by the quantity $s(L)$ defined as

$$s(L) = \inf p : ||L||_p < \infty . \tag{1.5}$$

With that we understand that, if $||L||_p < \infty$ for all real p, then $s(L) = -\infty$. In the case in which $||L||_p = \infty$ for all real p, we define $s(L) = \infty$. The quantity $s(L)$ will be called the <u>singular order</u> of the operator L.

If $s(L) = s$, where s is finite, then either

$$||L||_p \begin{cases} = \infty \text{ for } p < s, \\ < \infty \text{ for } p > s, \end{cases} \tag{1.6}$$

or

$$||L||_p \begin{cases} = \infty \text{ for } p < s, \\ < \infty \text{ for } p > s. \end{cases} \tag{1.7}$$

If we want to stress the difference, we shall way that the singular order of the operator L is "exactly s" in the first case, and "almost s" in the second case. By definition

$$\text{almost } s < \text{ exactly } s . \tag{1.8}$$

One can easily observe that, if $L = L' + L''$ and $s(L')>s(L'')$, then

$$s(L) = s(L') \tag{1.9}$$

If L is a (positive definite) differential operator of order <u>m</u>, then

$$s(L) = \text{exactly } \underline{m} + 3 . \tag{1.10}$$

Hence the singular order can be regarded as a generalization of the order of differential operators, shifted by 3 for convenience.

2. The convolution equations

Consider first the case in which the matrix function $\Lambda_{ij}(\underline{k})$ is a tempered distribution (i.e. all its components are tempered distributions). Then we have

$$\Lambda_{ij}(\underline{k}) = \hat{\Phi}_{ij}(\underline{k}) , \tag{2.1}$$

where $\hat{\Phi}_{ij}(\underline{k})$ is the Fourier transform of a tempered distribution $\hat{\Phi}_{ij}(\underline{x})$.

The fundamental equation can now be written in the convolution form

$$\Phi_{ij} * u_j = F_i , \tag{2.2}$$

with the kernel $\Phi_{ij}(\underline{x})$. The class $u_j(\underline{x})$, for which this equation is defined, still depends on a particular form of Φ_{ij}. In any case, we consider equation (2.2) equivalent to equation (3.10) with $\Lambda_{ij}(\underline{k})$ given by (4.1).

We have the following

Theorem (2.1). In order that $\Lambda_{ij}(\underline{k})$ be a tempered distribution it is necessary and sufficient that

$$s(L) < +\infty. \tag{2.3}$$

Proof: If $\Lambda_{ij}(\underline{k})$ is a tempered distribution, then, by (1.2), tr $\Lambda(\underline{k})$ is a positive tempered distribution. Such a distribution is given by a tempered measure, which implies inequality (1.4) for a certain real p. In consequence, we have inequality (2.3). To prove the converse, let us note that, the matrix $\Lambda_{ij}(\underline{k})$ being at least positive semi-definite for all $\underline{k}$, the inequality

$$|\Lambda_{ij}(\underline{k})| < 2 \text{ tr } \Lambda(\underline{k}) \tag{2.4}$$

holds for any pair of indices i , j. Thus, if the condition (2.3) is satisfied, then there exists a real p such that

$$\int d^3k(1+k^2)^{-p/2}|\Lambda_{ij}(\underline{k})|| < 2\,\|L\|_p < \infty \tag{2.5}$$

which shows that $\Lambda_{ij}(\underline{k})$ is a tempered distribution.

The singular order of a convolution equation provides a measure of the singularity of the kernel Φ_{ij}. The following theorems reveal the corresponding relation.

<u>Theorem (2.2)</u>. The kernel $\Phi_{ij}(\underline{x})$ is a continuous function if and only if

$$s(L) < \text{almost } 0\ . \tag{2.6}$$

Proof: If the inequality (2.6) is satisfied, then the inequality (2.5) holds for p = o - i.e. function $\Lambda_{ij}(\underline{k})$ is summable. By the Riemann-Lebesgue theorem, the kernel

$\Phi_{ij}(\underline{x})$, which is the Fourier transform of $\Delta_{ij}(\underline{k})$, is continuous. Conversely, if $\Phi_{ij}(\underline{x})$ is a continuous function, then

$$\text{tr } \Phi = \Phi_{11}(\underline{x}) + \Phi_{22}(\underline{x}) + \Phi_{33}(\underline{x}) \tag{2.7}$$

is a positive definite continuous function and, according to Bochner's theorem its Fourier transform is given by a finite measure.

This implies the inequality (1.4) for p = 0, and, in consequence, the condition (2.6).

Theorem (2.3). If

$$s(L) \leqslant \text{almost} - m \tag{2.8}$$

where m is a positive integer, then $\Phi_{ij}(\underline{x})$ has continuous derivatives up to order m.

Proof: Condition (2.8) implies the inequality (2.5) for $p = -m$. Taking into account the inequality

$$|k^{\mu}| \leqslant (1 + k^2)^{m/2} \quad \text{for} |\mu| = m \ , \tag{2.9}$$

we conclude that the derivatives $\partial^{\mu}\Phi_{ij}(\underline{x})$ with $|\mu| \leqslant m$ have summable Fourier transforms, and therefore, by the Riemann-Lebesgue theorem, are continuous.

Remark. For even integer m, the theorem converse to (2.3) is true. This can be shown by considering the kernel

$$\Phi'_{ij} = (1 - \Delta)^{m/2}\Phi_{ij} \ , \tag{2.10}$$

which defines an admissible operator L' with the matrix

$\Lambda_{ij}(\underline{k})$ given by

$$\Lambda_{ij}(\underline{k}) = (1 + k^2)^{m/2} \Lambda_{ij}(\underline{k}) \tag{2.11}$$

If the derivatives $\partial^{\mu}\Phi_{ij}$ are continuous for $|\mu| \leq m$, then the kernel Φ'_{ij} is continuous and, by Theorem (2.2),

$$s(L) = s(L') - m \leq \text{almost} - m \, . \tag{2.12}$$

From the remark and from Theorem (2.3), the following theorem follows immediately:

Theorem (2.4). The kernel $\Phi_{ij}(\underline{x})$ is an infinitely differentiable function, if and only if

$$s(L) = -\infty \tag{2.13}$$

Moreover, we have

Theorem (2.5). If tr $\Phi(\underline{x})$ is a bounded function in a certain neighbourhood of $\underline{x} = 0$, then the kernel $\Phi_{ij}(\underline{x})$ is a continuous function everywhere.

Proof: If assumption of the theorem is satisfied, then tr $\Phi(\underline{x})$ can be presented in the form

$$\text{tr}\ \Phi = f_1 + f_2 \, , \tag{2.14}$$

where f_1 is a bounded function,

$$|f_1(\underline{x})| < C \tag{2.15}$$

and f_2 is a tempered distribution such that

$$f_2(\underline{x}) = 0 \quad \text{for} \quad |\underline{x}| < \varepsilon \tag{2.16}$$

Where C and ε are certain positive constants. Let Ψ_α denote the function defined by the equation

$$\Psi_\alpha = \left(\frac{\alpha}{\sqrt{\pi}}\right)^3 e^{-\alpha^2 r^2}$$

for a certain value of the parameter α . Then

$$|(f_1, \Psi_\alpha)| < C \tag{2.17}$$

and there exists a polynomial $P(\alpha) > 0$ such that

$$|(f_2, \Psi_\alpha)| < e^{-\alpha^2 \varepsilon^2} P(\alpha) \tag{2.18}$$

Thus

$$\limsup_{\alpha\to\infty} |(\mathrm{tr}\ \Phi, \Psi_\alpha)| \leqslant C . \tag{2.19}$$

On the other hand,

$$(\mathrm{tr}\ \Phi, \Psi_\alpha) = (\mathrm{tr}\ \Lambda, \hat{\Psi}_\alpha) = \frac{1}{(2\pi)^3}\int d^3k\,\mathrm{tr}\Lambda\ (\underline{k})e^{-k^2/4\alpha^2} \tag{2.20}$$

and

$$\lim_{\alpha\to\infty} (\mathrm{tr}\ \Phi, \Psi_\alpha) = ||L||_o \leqslant C \tag{2.21}$$

where the last inequality follows from (2.19). Thus $s(L) \leqslant$ almost o and, by Theorem (2.2), the kernel $\Phi_{ij}(\underline{k})$ is continuous.

<u>Theorem (2.6)</u>. If

$$s(L) \leqslant \text{almost } m \quad , \tag{2.22}$$

where m is a positive integer, then the kernel $\Phi_{ij}(\underline{x})$ can be expressed as finite sum of continuous functions and their derivatives of order not greater than m.

Proof. We shall prove this theorem by the construction of the corresponding representation of the kernel Φ_{ij}.

Let

$$q = \begin{cases} \frac{m}{2} & \text{for even } m, \\ \frac{m+1}{2} & \text{for odd } m, \end{cases} \tag{2.23}$$

and

$$s(M) = s(L) + q < \text{almost } m - 2q. \tag{2.24}$$

Then $\Psi_{ij}(\underline{x})$ represents the kernel of an admissible convolution operator M of singular order

$$s(M) = s(L) + q \leqslant \text{almost } m - 2\,q \tag{2.25}$$

If m is even, then by Theorem (2.2) the kernel $\Psi_{ij}(\underline{x})$ is continuous and

$$\Phi_{ij} = (1 - \Delta)^q \Psi_{ij} \tag{2.26}$$

is a representation of the desired form. If m is odd, then by Theorem (2.3) the kernel Ψ_{ij} is continuously differentiable, so that Ψ_{ij} and

$$\chi_{lij} = \partial_l \Psi_{ij} \tag{2.27}$$

are continuous functions. Hence the representation we are looking for can be written as

$$\Phi_{ij} = (1 - \Delta)^{q-1} \partial_l \chi_{lij} + (1 - \Delta)^{q-1} \Psi_{ij} \tag{2.28}$$

It follows from Theorem (2.5) that whenever the singular

order of a convolution operator is negative (or exactly 0) i.e.

$$s(L) \leqslant \text{exactly } 0 \; , \tag{2.29}$$

then at $\underline{x} = 0$ the corresponding kernel has a singularity which cannot be represented by a bounded function.

This singularity, however, can be represented by derivatives of continuous functions, and Theorem (2.6) gives the dependence between the singular order of the operator and the necessary order of these derivatives.

3. The fundamental solution and the singular hardness of an elastic material

Consider now the inverse matrix $\Lambda^{-1}_{ij}(\underline{k})$, which is well defined at any $\underline{k} \neq 0$. Considered as a function of $\underline{k}$, this matrix is continuous for $k \neq 0$ and has a singularity at $\underline{k} = 0$.

However, as follows from the postulate III.G, this singularity is summable. Hence $\Lambda_{ij}(\underline{k})$ uniquely defines a locally summable function on the $\underline{k}$ - space, and it will be understood in this sense.

By the equation

$$\hat{u}_i(\underline{k}) = \Lambda^{-1}_{ij}(\underline{k}) \quad \hat{f}_j(\underline{k}) \quad , \tag{3.1}$$

the function $\Lambda^{-1}_{ij}(\underline{k})$ defines an operator from L[U] into U which will be denoted by L^{-1}:

$$L^{-1}_{ij} f_j(\underline{x}) = u_i(\underline{x}) \quad . \tag{3.2}$$

In the case in which $\Lambda^{-1}_{ij}(\underline{k})$ is a tempered distribution (and this depends solely on its asymptotic bahaviour for $k \to \infty$), there exists a tempered distribution $G_{ij}(\underline{x})$ whose Fourier transform

$$\hat{G}_{ij}(\underline{k}) = \Lambda^{-1}_{ij}(\underline{k}) \quad . \tag{3.3}$$

In this case, the distribution $G_{ij}(\underline{x})$ satisfies the equation

$$L_{ij} G_{jm}(\underline{x}) = \delta_{im} \delta^{(3)}(\underline{x}) \tag{3.4}$$

and will be called the fundamental solution.

The function $\Lambda^{-1}_{ij}(\underline{k})$ being locally integrable, the definitions (1.1) and (1.5) make sense for the operator L^{-1}. Moreover, all the theorems of section 2 apply to the operator L^{-1}, provided that the following substitutions are made:

$$\begin{aligned} \Phi_{ij} &\longrightarrow G_{ij}, \\ \Lambda_{ij} &\longrightarrow \Lambda^{-1}_{ij}, \\ s(L) &\longrightarrow s(L^{-1}). \end{aligned} \tag{3.5}$$

In particular, the fundamental solution

a) exists,

b) is continuous (bounded),

c) is infinitely differentiable,

if and only if

a) $s(L^{-1}) < +\infty$,

b) $s(L^{-1}) \leqslant$ almost 0, (3.6)

c) $s(L^{-1}) = -\infty$

respectively.

The quantity

$$h(L) = -s(L^{-1}) \tag{3.7}$$

will be called the <u>singular hardness</u> of the corresponding elastic material.

The idea here is that if the material is "singular hard", the singularity of the displacement field created by applying a concentrated force is weak. And if the material is "singular soft", a concentrated force creates a strong singularity in the displacement field. The quantity (3.7) provides a numerical measure of this property.

4. The relation between s(L) and $s(L^{-1})$

Now we shall prove the following fundamental inequality between the singular orders of the operators L and L^{-1}:

$$s(L^{-1}) \geqslant 6 - s(L) . \tag{4.1}$$

The proof is based on the following inequality:

$$\int tr(A^2)(1+k^2)^{-p/2}d^3k \int tr(3^2)(1+k^2)^{-p'/2}d^3k \geqslant$$
$$\geqslant (\int |tr\ AB|(1+k^2)^{\frac{-p-p'}{4}} d^3k)^2 , \tag{4.2}$$

which can be obtained from Schwartz's inequalities for traces and integrals and is valid for any measurable matrix functions $A_{ij}(\underline{k})$, $B_{ij}(\underline{k})$ and arbitrary real numbers p, p'.

By substituting $\Lambda = A^{1/2}$ and $B = \Lambda^{-1/2}$ into (4.2) we obtain

$$||L||_p \; ||L^{-1}||_{p'} \geqslant 9||1||^2_{\frac{p+p'}{2}} \quad . \tag{4.3}$$

Thus, whenever

$$||L||_p < \infty \text{ and } ||L^{-1}||_{p'} < \infty \tag{4.4}$$

then

$$p' > 6 - p \, , \tag{4.5}$$

the right-hand side of the inequality (4.3) being devergent on the contrary. Hence

$$\inf p' \geqslant \sup (6 - p) = 6 - \inf p \, , \tag{4.6}$$

which proves the inequality (4.1).

By introducing the following convention:

$$- \text{almost } s = \text{exactly} - s \tag{4.7}$$

the validity of (4.1) is extended to singular orders labelled as "almost" or"exactly" .

Corollary (4.1). For the fundamental solution G_{ij} to be bounded (continuous), it is necessary that

$$s(L) \geqslant \text{exactly } 6 \; . \tag{4.8}$$

Corollary (4.2). If the kernel Φ_{ij} is bounded (continuous), then

$$s(L^{-1}) > \text{exactly } 6 \ . \tag{4.9}$$

<u>Corollary (4.3)</u>. If the kernel Φ_{ij} is infinitely differentiable, then the fundamental solution does not exist.

The last corollary explains the failure of the example discussed in Chapter I: the kernel in this example is infinitely differentiable.

Generally, the smoother the kernel Φ_{ij} of a convolution equation is, the more singular the fundamental solution must be. The weakest possible singularity of G_{ij} corresponds to the value of $s(L^{-1})$ given by equation

$$s(L^{-1}) = 6 - s(L) \tag{4.10}$$

The following two examples illustrate how inequality (4.1) works in some particular cases.

1. Consider an isotropic medium described by a convolution fundamental equation. In that case, the general form of the kernel $\Phi_{ij}(\underline{x})$ is

$$\Phi_{ij}(\underline{x}) = -(\Delta\delta_{ij} - \partial_i\partial_j)\Psi_1(r) - \partial_i\partial_j\Psi_2(r) \tag{4.11}$$

where $\Psi_1(r)$ and $\Psi_2(r)$ are spherically symmetric tempered distributions.

Let $\Psi_1(r)$ and $\Psi_2(r)$ have singularities at $r=0$ only, and let

$$\Psi_1(r) \sim r^{2-\alpha_1}, \ \Psi_2(r) \sim r^{2-\alpha_2}, \text{ for } r\to 0 \tag{4.12}$$

with some non-integer α_1 and α_2. Then the singular order of the corresponding operator L equals

$$s(L) = \text{exactly } \max(\alpha_1, \alpha_2) \ . \tag{4.13}$$

The fundamental solution has the same form:

$$G_{ij}(\underline{x}) = -(\Delta\delta_{ij} - \partial_i\partial_j)H_1(r) - \partial_i\partial_j H_2(r) \tag{4.14}$$

and, as inspection of the corresponding Fourier transforms shows, the strongest singularity is again at r = o, and

$$H_1(r) \sim r^{\alpha_1 - 4}, \ H_2(r) \sim r^{\alpha_2 - 4}, \quad \text{for } r \to 0 \ . \tag{4.15}$$

Hence

$$S(L^{-1}) = \text{exactly } 6 - \min(\alpha_1, \alpha_2) \ . \tag{4.16}$$

Thus in the present example equation (4.10) holds numerically if $\alpha_1 = \alpha_2$. On the contrary, the sharp numerical inequality in (4.10) is valid.

2. Consider an elastic medium described by nonlocal stress-strain relation of the following form:

$$6_{ij}(\underline{x}) = c_{ijkl}\varepsilon_{kl}(\underline{x}) + \int c^*_{ijkl}(\underline{x}-\underline{x})\varepsilon_{kl}(\underline{x}')d^3x' \ . \tag{4.17}$$

The corresponding fundamental equation has the convolution form (2.2) with the kernel

$$\Phi_{ij}(\underline{x}) = -\partial_k\partial_l(c_{ijkl}\delta^{(3)}(\underline{x}) + c^*_{ijkl}(\underline{x}) \ ) \tag{4.18}$$

Let the function $c^*_{ijkl}(\underline{x})$ be absolutely integrable . Then, by writing the corresponding Fourier transforms and making use of the Riemann-Lebesgue theorem, we obtain

$$s(L) = \text{exactly } 5 \ . \tag{4.19}$$

Hence, according to Corollary (4.1), the fundamental solution cannot be bounded or continuous. In fact, an inspection of relevant Fourier transforms shows that

$$s(L^{-1}) = \text{exactly } 1 \ . \tag{4.20}$$

Thus, in the case considered equation (4.10) is numerically valid. This also refers to the classical case ($c^*_{ijkl} = 0$).

5. Non-convolution equations

In section 2, we have discussed the case in which $\Lambda_{ij}(\underline{k})$ is a tempered distribution. Now, we shall consider the remaining case. If, $\Lambda_{ij}(\underline{k})$ is not tempered distribution, then there is no distribution, even non-tempered, which would allow the fundamental equation to be written in the convolution form (2.2). Nevertheless, such a function $\Lambda_{ij}(\underline{k})$ defines uniquely an operator L, and this operator has all the properties required. We shall refer to this case as the non-convolution case. The corresponding singular order of L is

$$s(L) = +\infty \ . \tag{5.1}$$

In this case, the inequality (4.1) does not restrict the regularity of the fundamental solution, which can be an infinitely differentiable function.

In fact, we have

<u>Proposition (5.1)</u>. If for any real m the inequalities

$$\omega_1^2(\underline{k}) > k^m, \quad \omega_2^2(\underline{k}) > k^m, \quad \omega_3^2(\underline{k}) > k^m \tag{5.2}$$

are satisfied provided that the vector $\underline{k}$ is sufficiently large, then

$$s(L) = +\infty, \qquad s(L^{-1}) = -\infty . \tag{5.3}$$

This proposition follows directly from the definition of $s(L)$ and $s(L^{-1})$.

Consider an example. Let

$$\Lambda_{ij}(\underline{k}) = [\mu k^2 \delta_{ij} + (\lambda + \mu) k_i k_j] e^{k^2/4\beta^2} . \tag{5.4}$$

The Fourier transform of the fundamental solution is

$$\hat{G}_{ij}(\underline{k}) = [\frac{1}{\mu}\frac{\delta_{ij}}{k^2} - \frac{k_i k_j}{k^4} + \frac{1}{\lambda+2\mu}\frac{k_i k_j}{k^4}] e^{-k^2/4\beta^2} . \tag{5.5}$$

Hence the fundamental solution equals

$$G_{ij}(\underline{x}) = \frac{1}{4\pi\mu}\delta_{ij}\frac{\mathrm{erf}(\beta r)}{r} +$$

$$-\frac{1}{4\pi}\left(\frac{1}{\mu} - \frac{1}{\lambda+2\mu}\right)\partial_i \partial_j [\frac{1}{r}\int_0^r (r-s)\,\mathrm{erf}(\beta s)\,ds] . \tag{5.6}$$

This solution is not only infinitely differentiable: it is an entire analytic function of $\underline{x}$.

VII. GEOMETRICAL AND DYNAMICAL NON-LOCALITY

1. Basic concepts

The need to describe the physical reality in a more adequate way gives rise to various modifications of the classical theories. The modified operators do not necessarily fulfill the locality conditions.

To define an operator A means (i) to define the domain Y of A and (ii) to specify the action of A in Y. Hence, to modify an operator A means either (i) to modify the domain Y, or (ii) to modify the action A, or else both. This opens two general ways of modifying the classical theories.

The first of them, the geometrical (or kinematical) one, consists in making a substantial modification of the domain. Without changing the physical laws of interaction of particles, the set of admissible configurations (or motions) is controlled.

The second way, which can be called dynamical, consists in modifying the form of the operator A without making any substantial changes of the domain. In terms of material continua it corresponds to a modification of interactions of material particles.

Although the dynamical approach is nowadays most popular, it does not mean that the geometrical one is not interesting. The impression that the physical contents of a theory can not be seriously changed by modifying the set of admissible functions is due to considering not too significant modifications of Y. The pseudo-continuum theory [Ch.V] furnishes an example of the contrary. Generally, from the atomistic point of view one can argue that the set of functions admissible in the classical continuum theory is definitely too large. In fact, many unphysical results of this theory are directly due to excessive freedom of the classical continuum in forming singularities, viabrations of very high frequency, etc. Therefore, the idea of forming the domain Y by a substantial restriction of the classical resources of functions seems to be promising.

The Definition I. (3.1) expresses the intuitive idea of locality of an operator provided that Y contains functions of arbitrarily small compact support, concentrated around an arbitrary point $x \in \Omega$. If Y does not contain functions of compact support at all, then there is no difference between local and nonlocal operators, although the conditions I. (3.2) and I. (3.3) are trivially satisfied.

Bearing that in mind, we shall introduce the following definitions.

Definition 1.1 . A space Y of functions over a domain Ω is called local if for any $x \in \Omega$ and for any neighbourhood N such that $\Omega \supset N \ni x$, there exists $y \in Y$ which satisfies

$$x \in \text{supp } y \subset N \tag{1.4}$$

Definition 1.2 . A theory based on a fundamental equation of the form (1.1) is local if Y is a local space and A is a local operator.

A theory which is not local will be considered to be nonlocal. From Definition 1.2. it follows that there are two basic reasons for the nonlocality of a theory: either non-locality of the space Y or nonlocality of the operator A. These two cases will be reffered to as geometrical and dynamical nonlocality, respectively.

2. The spaces Y

To avoid considering too poor or too irregular spaces Y let us make the following assumptions.

Let Ω be an open region in a Euclidean space E, $\bar{\Omega}$ - its closure. Let C^{∞} denote the space of functions on E which have continuous derivatives of arbitrary order. In the case of Ω with a boundary, by C^{∞} we shall understand the space of restrictions of these functions from E to $\bar{\Omega}$. If it is necessary to indicate the region, we shall write explicitly $C^{\infty}(\bar{\Omega})$ or $C^{\infty}(E)$. The same applies to other functions spaces, such as C_o^{∞} which consists of infinitely differentiable functions of compact support, or C_b^{∞} which is composed of bounded C^{∞}- functions.

The spaces Y will be considered as configuration spaces of certain physical systems, although other interpretations are also possible. In this paper we shall not profit much from the conceptual difference between a physical system and its configuaration space. Thus we shall treat the corresponding expressions as synonyms.

The spaces Y will be assumed to be linear, and endowed with an appriopriate topology when necessary.

Definition 2.1 . A system is called classical continuum if

$$C_o^{\infty} \subset Y \quad . \tag{2.1}$$

A classical theory is always geometrically local.

Definition 2.2 . A system is called non-classical continuum if (i) there exists a function $\ell \in C_o^{\infty}$ which does not belong to Y, and (ii) Y contains all the functions of the form

$$e^{ikx} \ell(x) \tag{2.2}$$

(or their real counterparts) with arbitrary real k and $\ell(x) \in Y$.

As to geometrical locality or non-locality, both occur in non-classical continua.

Definition 2.3 . A system is called pseudo-continuum if there exists a compact set D in the k-space such that for any $y \in Y$

$$\mathrm{supp}\,\hat{y} \subset D \,, \tag{2.3}$$

where $\hat{y}$ denotes the Fourier transform of y (or, more generally, of a certain extension of y from $\bar{\Omega}$ to E).

Pseudo-continua are always nonlocal.

We shall not be interested in functions which grow up very fast when $|x| \to \infty$. We shall consider (i) functions of tempered growth, i.e. such that there exist C and N (dependent on ℓ and μ) satisfying the inequality

$$|\partial^{\mu} \ell(x)| \leqslant C(1 + |x|^{N}) \quad , \tag{2.4}$$

(ii) functions which are bounded together with all their derivatives,

$$|\partial^{\mu} \ell(x)| \leqslant \mathrm{const} \,, \tag{2.5}$$

where the constant can depend on μ , and (iii) functions of fast decrease, i.e. such that for any μ and N there exists C satisfying the inequality

$$|\partial^{\mu} \ell(x)| \leqslant C(1 + |x|^{-N}) \quad . \tag{2.6}$$

3. Construction of non-classical continua

In this section we shall describe a spectrum of nonclassical continua. In order to concentrate our attention on the basics facts, we shall avoid discussing multidimensional cases, restricting ourselves to one-domensional E. The qth order derivative of ℓ wil be denoted by $\ell^{(q)}$.

Let us start from the following observation. Let $\{b_q\}$ be an arbitrary positive sequence,

$$b_q > 0 \qquad \text{for} \quad q = 0, 1, \ldots, \tag{3.1}$$

and x - and arbitrary point in E. We shall say that a function $\ell \varepsilon C^{\infty}$ is majorized at x by the sequence $\{b_q\}$ if there exists a constant C such that

$$|\ell^{(q)}(x)| < Cb_q \tag{3.2}$$

for any $q > 0$. With this definition one can assert that for any $\{b_q\}$ there exists a function $\ell \varepsilon C_o^{\infty}$ which is not majorized by $\{b_q\}$. A proof of this assertion can be given by constructing such a function, and will be not given here. Informally speaking the sequence of consecutive derivatives of an infinitely differentiable function (C_o^{∞} and, a fortiori, C^{∞}) can grow up arbitrarily fast.

The above obseravtion suggests the idea of constructing non-classical continua by making use of the inequalities (3.2) with appropriate classes of sequences $\{b_q\}$. In general, the constant C in the inequalities (3.2) can be dependent on x. This dependence, although insignifficant on compact Ω's, can be made use of for defining the behaviour of admissible functions at infinity.

We shall be interested in functions which grow up not faster than polynomials when $|x| \to \infty$.

Which sequences $\{b_q\}$ are interesting from the point of view of constructing non-classical continua?

Let us consider first the $\{b_q\}$'s which grow up more slowly than any power sequence, i.e. for any $B > 0$ there exists C such that

$$b_q \leqslant CB^q \tag{3.3}$$

Then ℓ is an entire analytic function such that for any z and B

$$|\ell(x + z)| \leqslant Ce^{|B|z} , \tag{3.4}$$

which means that it is either of the fractional order of growth or of minimal exponential type. In both cases it follows that all the ℓ's which are not polynomial have to grow up faster than any polynomial when $x \to \infty$ or $x \to -\infty$ (i.e. at least in one real direction).

Therefore we shall restrict ourselves to sequences $\{b_q\}$ which grow up like power sequences or faster. Any such sequence can be represented in the form

$$b_q = d_q B^q \tag{3.5}$$

where d_q is a non-decreasing sequence.

Consider the sequences which satisy the inequality (3.3) with a finite B. Then, from the inequality (3.4) it follows that ℓ is an entire analytic function of a finite exponential type. There exist functions of this type and of moderate growth for $|x|\to\infty$. By the Paley-Wiener-Schawartz theorem any such function has the Fourier transform of compact support.

Hence sequence $\{b_q\}$ satisfying the inequality (3.3) with a finite B define pseudocontinua in the sense of the Definition 2.3.

Consider now the sequences $\{b_q\}$ of faster growth, i.e. of the form (3.5) with increasing d_q ,

$$\limsup d_q = \infty. \tag{3.6}$$

Let

$$\sum_{q=0}^{\infty} c_q z^q \tag{3.7}$$

where

$$c_q = -\frac{1}{q!}\ell^{(q)} \tag{3.8}$$

are the coefficients of Taylor's expansion of ℓ at a certain point x. According to basic theorems on analytic functions, the radius of convergence of this series equals

$$R = \liminf |c_q|^{-1/q} . \tag{3.9}$$

If $R = \infty$, ℓ is an entire analytic function of the order given by

$$\rho = \lim\sup \frac{q\ln q}{\ln \frac{1}{|c_q|}} \quad . \tag{3.10}$$

Appriopriate calculations show that if ℓ satisfies the inequality (3.2) then

$$\ln\frac{1}{R} \leqslant 1 + \lim\sup(\frac{1}{q}\ln b_q - \ln q) \ , \tag{3.11}$$

$$\frac{1}{\rho} \geqslant 1 - \lim\sup\frac{\ln b_q}{q\ln q} \quad . \tag{3.12}$$

By substituting eqn (3.5) into the inequality (3.12) we obtain

$$\ln\frac{1}{R} \leqslant 1 + B + \lim\sup\frac{1}{q}\ln d_q - \ln q \ , \tag{3.13}$$

$$\frac{1}{\rho} \geqslant 1 - \lim\sup\frac{\ln d_q}{q\ln q} \quad . \tag{3.14}$$

These formulae suggest the following chice of $\{d_q\}$'s:

$$\frac{\ln d_q}{q\ln q} = \beta \tag{3.15}$$

which gives

$$d_q = q^{\beta q} \quad . \tag{3.16}$$

In order to obtain increasing sequences $\{d_q\}$, we must assume $\beta > 0$.

Now we can define the following family of function spaces.

Definition 3.1. The space $Q^{\beta,B}(E)$ consists of all the C^{∞}-functions of moderate growth such that

$$|\ell^{(q)}(x)| \leq Cq^{q}\,\bar{B}^{q} \tag{3.17}$$

for a certain $C = C(x)$ of moderate growth and any $\bar{B} > B$ (i.e. $\Lambda\bar{B}\ C\Lambda q$).

Extending the above definition by admitting $\beta = 0$ allows us to include also quasicontinuum spaces $Q^{0,B}$.

By substituting into Definition 3.1 phrases like "of bounded derivatives" or "of fast decrease" in place of the phrase "of moderate growth", one can obtain valid definitions for the corresponding spaces with different bahaviour of functions at infinity.

Definition 3.2. The space $Q^{\beta}(E)$ is the union of all spaces $Q^{\beta,B}(E)$.

Definition 3.3. The spaces $Q^{\beta}(\bar{\Omega})$ and $Q^{\beta,B}(\bar{\Omega})$ are composed of corresponding restrictions of functions from $Q^{\beta}(E)$ and $Q^{\beta,B}(E)$, respectively.

The spaces $Q^{\beta,B}$ and Q^{β} will be briefly reffered to as Q-spaces.

4. Basic properties of Q-spaces

If either $\beta' < \beta$ or $\beta' = \beta$ and $B' < B$, then the following inclusion

$$Q^{\beta',B'} \subset Q^{\beta,B} \tag{4.1}$$

holds and is proper. It follows from the fact that the space $Q^{\beta,B}$ contains $S^{\beta,B}_{\alpha,A}$, where $S^{\beta,B}_{\alpha,A}$ are the spaces of C^{∞}-functions satisfying the inequalities

$$|x^{k}{}_{\ell}^{(q)}(x)| \leqslant Cq^{\beta q}\bar{B}^{q}k^{\alpha k}\bar{A}^{k} \tag{4.2}$$

for a certain C and any $\bar{B} > B$, $\bar{A} > A$. The parameters and A in the relation (4.1) satisfy $\alpha > 0$, $A > 0$, and otherwise are arbitrary. As the spaces $S^{\beta,B}_{\alpha,A}$ with $\alpha+\beta > 1$ are non-empty, so are all the spaces $Q^{\beta,B}$. This conclusion can be strengthened a little by taking into account the fact that $S^{\beta,B}_{\alpha,A}$ contain only functions of fast decrease.

By putting eqn (3.16) into eqns (3.13) and (3.14) one obtains

$$\begin{aligned} R &= 0 && \text{for} \quad \beta > 1, \\ R &\geqslant \frac{1}{eB} && \text{for} \quad \beta = 1, \\ R &= \infty && \text{for} \quad \beta < 1 \end{aligned} \tag{4.3}$$

and

$$\rho \leqslant \frac{1}{1-\beta} \quad . \tag{4.4}$$

Hence:

If $\beta > 1$, then $Q^{\beta,B}$ - functions are not analytic;

If $\beta = 1$, then $Q^{\beta,B}$ - functions are either analytic with a finite radius of convergence satisfying eqn $(4.3)_2$, or entire analytic functions of infinite order;

If $\beta < 1$, then $Q^{\beta,B}$ - functions are entire analytic of finite order given by eqn (4.4). When β varies from 1 to 0, the order ρ varies from ∞ to 1.

Since the functions belonging to $Q^{\beta,B}$ with $\beta < 1$ are analytic (at least of a finite radius at any x), these spaces do not contain functions of compact support and, in consequence, are not local.

Making use of the fact that $S^{\beta,B}_{0,A}$ consists of functions which vanish identically for $|x| > A$, and taking $\alpha = 0$, $\beta > 1$ in eqn (4.1), one concludes that for $\beta > 1$ any $Q^{\beta,B}$ contains C_0^∞-functions of arbitrarily small supports. Hence, according to Definition 1.1, the spaces $Q^{\beta,B}$ with $\beta > 1$ are local.

5. Basic operations in Q-spaces

The spaces $Q^{\beta,B}$ are linear. Consider the following $\ell \to \Psi$ operations:

(i) translation by arbitrary real a

$$\Psi(x) = \ell(x - a) , \tag{5.1}$$

(ii) differentiation

$$\Psi(x) = \ell'(x) . \tag{5.2}$$

Proposition 5.1. The operations of translation and differentiation are

$$Q^{\beta,B} \to Q^{\beta,B} \tag{5.3}$$

(i.e. defined on $Q^{\beta,B}$ and having values in $Q^{\beta,B}$) for all β, B.

A function f is called multiplier in $Q^{\beta,B}$ if the multiplication operation

$$\ell \to f\ell \quad (5.4)$$

is (5.3) .

Proposition 5.2. Let $\beta' < \beta$ and

$$f \in Q^{\beta',B} \quad (5.5)$$

with an arbitrary B'. Then f is a multiplier in $Q^{\beta,B}$.

Proposition 5.3. Let the relation (5.5) hold. Then the multiplication operation (5.4) is

$$Q^{\beta',B} \to Q^{\beta',B+B} \quad . \quad (5.6)$$

The proof will be omitted here.

Corollary 5.1. Polynomials are multipliers in all spaces $Q^{\beta,B}$.

Corollary 5.2. The functions exp(ikx) with arbitrary real k are multipliers in any space $Q^{\beta,B}$ with $\beta > 0$.

Proof. Follows from Proposition 5.2. since the functions exp(ikx) belong to $Q^{0,k}$.

It follows from this Corollary that all the spaces $Q^{\beta,B}$ with $\beta > 0$ are non-classical continua.

Consider a scale transformation operation defined by the equation

$$\Psi(x) = \ell(\lambda x) \ , \qquad \lambda > 0 \quad (5.7)$$

Proposition 5.4. The scale transformation (5.7) is

$$Q^{\beta,B} \to Q^{\beta,\lambda B} \quad . \quad (5.8)$$

Corollary 5.3. The scale transformation operation (5.7) is

$$Q^{\beta} \to Q^{\beta} . \quad (5.9)$$

Hence the spaces Q^{β} are invariant with repsect to the scale transformations, while $Q^{\beta,B}$ are not.

6. Differential operators of infinite order. Summary

A brief summary of Q-spaces and the corresponding terminology is given in Table 1. The last column indicates another interesting property of non-classical continua: while every differential operator in a classical continuum has to be of finite order, non-classical continua admit linear differential operators of infinite order. The conclusions follow from the following simple considerations.

Let

$$f(\partial) = \sum_{q}^{\infty} f_q \partial^q \quad (6.1)$$

be a formal differential operator of (possibly) infinite order. Then the action of eqn (6.1) on a C^{∞}-function should be given by

$$f(\partial)\ell = \sum_{q}^{\infty} f_q \ell^{(q)} . \quad (6.2)$$

According to the observation formulated in the Section 3 for any $\{f_q\}$ there exists $\ell \in C_0^{\infty}$ such that eqn (6.2) is divergent, exept the case of

$$f_q = 0 \quad \text{for every} \quad q > q_0 . \tag{6.3}$$

On the other hand, all ℓ's belonging to a non-classical Q-space satisfy eqn (3.17). In consequence, for any Q-space there exists infinite $\{f_q\}$'s which make eqn (6.2) convergent.

All admissible differential operators are local, and this locality is essential in local spaces. In nonlocal spaces every linear operator can be equivalently expressed in a differential form of order $\leqslant \infty$.

Table 1

Q-spaces	$\beta = \infty$	$\infty > \beta > 1$	$\beta = 1$	$1 > \beta > 0$	$\beta = 0$
terminology	classical	non-classical			
	local		nonlocal		
	continuum				pseudo-continuum
functions admitted	C^∞	restricted C^∞	analytic	entire analytic	finite exponential type
linear differential operators	finite order	infinite order	—	—	—

BIBLIOGRAPHY

Hutchins, R.M., Ed., *The works of Aristotle I, II*, W. Branton Publ., 1962.

Cosserat, E., Cosserat, F., *Theorie des corps deformables*, Hermann 1909.

Edelen, D.G.B., *Nonlocal variations and local invariance of fields*, American Elsevier, New York, 1969.

Edelen, D.G.B., *Nonlocal variational mechanics*, IJES **7**, 269, 287, 373, 391, 401, 677, 843 (1969), 517 (1970); **13**, 861, (1975).

Edelen, D.G.B., *Invariance theory for nonlocal variational principles*, IJES **9**, 741, 801, 819, 921, (1971).

Edelen, D.G.B., *A nonlocal variational formulation of the equations of radiative transport*, IJES **11**, 1109, 1973.

Edelen, D.G.B., *Irreversible thermodynamics of nonlocal systems*, IJES **12**, 607, 1974.

Edelen, D.G.B., *On compatibility conditions and stress boundary value problems in linear nonlocal elasticity*, IJES **13**, 971, 1975.

Edelen, D.G.B., *Theories with carrier fields: multiple interaction nonlocal formulation*, Arch. Mech. **28**, 1976, 353.

Edelen, D.G.B., *Nonlocal field theories in: continuum physics*, Academic Press 1976.

Edelen, D.G.B., *A global formulation of continuum physics and the resulting equivalence classes of nonlocal field equations*, in: Nonlocal theory of material systems, Ed. D. Rogula, Ossolineum, Wrocław, 1976.

Edelen, D.G.B., Green, A.E., Laws, N., *Nonlocal continuum mechanics*,, ARMA **43**, 36, 1971.

Eringen, A.C., Edelen, D.G.B., *On nonlocal elasticity*, IJES**10**, 233, 1972.

Eringen, A.C., *Nonlocal polar elastic continua*, IJES **10**, 1, 1972.

Eringen, A.C., *Linear theory of nonlocal elasticity and dispersion of plane waves*, IJES**10**, 425, 1972.

Eringen, A.C., *Theory of nonlocal electromagnetic elastic solids*, J. Math. Phys. **14**, 733, 1973.

Eringen, A.C., *Theory of nonlocal thermoelasticity*, IJES **12**, 1063, 1974.

Eringen, A.C., Kim, B.S., *On the problem of crack tip in nonlocal elasticity*, in: Continuum Mechanics aspects of Geodynamics and Rock Fracture, 1974.

Datta Gairola, B.K., Kröner, E., *The nonlocal theory of elasticity and its application to interaction of point defects*, *in:* Nonlocal theory of material systems, Ed.: D. Rogula, Ossolineum, Wrocław, 1976.

Gairola, B.K.D., *The nonlocal theory of elasticity and its application to interaction of points defects*, Arch. Mech. **28**, 393, 1976.

Holnicki-Szulc, J., Rogula, D., *Nonlocal continuum models of large engineering structures*, Arch. Mech **31**, 793, 1979.

Holnicki-Szulc, J., Rogula, D., *Boundary problems in nonlocal continuum models of large engineering structures*, Arch. Mech. **31**, 803, 1979.

Kapelewski, J., Rogula, D., *Pseudocontinuum approach to the theory of interactions between impurity defects and crystal lattice*, Arch. Mech. **31**, 27, 1979.

Kotowski, R., Rogula, D., *Differential pseudocontinua*, Arch. Mech., **31**, 43, 1979.

Kröner, E., Datta, B.K., *Nichtlokale Elastostatik: Ableitung aus der Gittertheorie*, Z.F. Physic, **196**, 203, 1966.

Kröner, E., *Elasticity theory of materials with long-range cohesive forces*, IJSS **3**, 731, 1967.

Kröner, E., *The problem of non-locality in the mechanics of solids: review of present status, in:* Fundamental aspects of dislocation theory, Eds.: J.A. Simmon, NBS Special Publ. II 1970.

Krumhansl, J.A., *Generalized continuum field representation for lattice vibrations*, in: Lattice dynamics, Ed. E. Wallis, Pergamon Press, 1965.

Kunin, A., *Model of elastic medium simple structure with spatial dispersion*, Prikl. Math. Mech. **30**, 942, 1966, in Russian.

Kunin, I.A., *Theory of elasticity with spatial dispersion. One-dimensional complex structure*, Prikl. Math. Mech. **30**, 866, 1966 in Russian.

Kunin, I.A., *Inhomogeneous elastic medium with nonlocal interaction*, Prikl. Mat. Tech. Phys. **3**, 60, 1967.

Kunin, I.A., *Theory of elastic media with microstructure. Nonlocal theory of elasticity*, Moscow 1975, in Russian.

Kunin, I.A., Vaisman, A.M., *On problems of the nonlocal theory of elasticity* in: Fundamental aspects of dislocation theory, NBS Spec. Publ. 1970.

Rogula, D., *Influence of spatial acoustic dispersion on dynamical properties of dislocation, I, II*, Bull, Acad. Pol. Sci. ser. tech. **13**, 337, 1965.

Rogula, D., *On nonlocal continuum theories of elasticity*, Arch. Mech. **25**, 233, 1973.

Rogula, D., *Some basic solutions in strain-gradient elasticity of an arbitrary order*, Arch. Mech. **25**, 43, 1973.

Rogula, D., *Dislocation lines in nonlocal elastic continua*, Arch. Mech. **25**, 967, 1973.

Rogula, D., *Quasicontinuum theory of crystals*, Arch. Mech. **28**, 563, 1976.

Rogula, D., *Geometrical and dynamical nonlocality*, Arch. Mech. **31**, 65, 1979.

Rogula, D., *Generalized interactions in nonlocal continua*, in: Continuum models of discrete systems, University of Waterloo Press 1980.

Rogula, D., Sztyren, M., *On the one-dimensional models on nonlocal elasticity*, Bull, Acad. Pol. Sci., ser. tech. **26**, 341, 1978.

Rogula, D., Sztyren, M., *Fundamental one-dimensional solutions in nonlocal elasticity*, Bull. Acad. Pol. Sci., ser. tech. **26**, **417**, **1978**.

Rogula, D., Kotowski, R., *On the correspondence between representations in the pseudocontinuum theory*, Bull. Acad. Pol. Sci., ser. tech. **26**, 555, 1978.

Russel, B., *The scientific outlook*, G. Allen and Unwin, London 1954.

Rymarz, Cz., *Continuous nonlocal models of a bounded elastic medium*, Proc. Vibr. Probl., **15**, 283, 1974.

Rymarz, Cz., *Boundary problems on the nonlocal theory*, Proc. Vibr. Probl. **15**, 355, 1974.

Rzewuski, J., *Field theory II*, Iliffe Books Ltd., London, PWN Warsaw 1969.

Sztyren, M., *On the boundary forces in a solvable integral model of nonlocal elastic half-space*, Bull. Acad. Pol. Sci., ser. tech. **26**, 537, 1978.

Sztyren, M., *Boundary value problems and surface forces for models of nonlocal elastic bodies*, Bull. Acad. Pol. Sci., ser. tech. **27**, 1979.

Sztyren, M., *Boundary value problems and surface forces for integro-differential models of nonlocal elastic bodies*, Bull. Acad. Pol. Sci., ser. tech. **27**, 1979.

Sztyren, M., *On nonlocal boundary value problems*, in: CMDS3 Proceedings, Univ. of Waterloo Press 1980.

Woźniak, Cz., *On the nonlocal effects in continuum mechanics due to internal constraints*, in: Nonlocal theory of material systems, Ed.: D. Rogula, Ossolineum, Wrocław 1976.

Zorski, H., Rogula, D., Rymarz, Cz., *Nonlocal continuum models of discrete systems*, Advances in Mechanics, **1**, 1979, in Russian.

ON SOLVABLE NONLOCAL BOUNDARY-VALUE PROBLEMS

Małgorzata Sztyren

Technical University of Warsaw, Institute of Mathematics

Plac Jednosci Robotniczej 1, 00-661 Warsaw, Poland

Introduction

The nonlocal models which are being used in practice can be classified with respect to the form of their static equations into three categories: the volume-integral (VIM), the volume-surface integral (VSIM), and the integro-differential models (IDM). They will be discussed in more detail later. The general theory of nonlocal models for linear, homogeneous elastic media given by Rogula[1] permits a systematic approach to the construction of such models. In this paper we shall not construct of them but only discuss the existing actually used class of models.

Mathematical features of nonlocal models have recently been investigated thoroughly. The results reveal that the physical properties of certain nonlocal models are not always satisfactory. In particular,solutions with desired properties which would assure a sufficient description of the studied phenomena, do not exist.

In present course of lectures we shall examine the three types of models mentioned above. To some extent unbounded media will be considered, and the existence and properties of fundamental solutions will be investigated. But specially we shall be interested in nonlocal bodies with boundaries. From the purely mathematical point of view nonlocal boundary-value problems for elliptic partial differential operators were considered by Beals[2]. Boundary-value problems for nonlocal models of material media have been studied by Weissman and Kunin[3], Rymarz[4], Sztyren[5-7].

In the present paper we shall examine the relation between boundary-value problems and near-surface forces. In what follows, by near boundary force we shall understand a volume distribution of force concentrated in a thin near-surface layer. The term surface force will be reserved for a distribution of force concentrated precisely at the surface.

The main subject of our analysis will be the case of external loadings in the form of near-boundary forces. We raise the question when, if the thickness of the layer corresponding to the effective range of the near-boundary forces tends to zero, the limit transition exists and whether the limit solution (if it exists) equals the solution for the corresponding surface force. This problem is essential since in nature the "contact" forces are always a little diffuse.

A solvable example of a nonlocal boundary-value problem

Let us consider an example of a nonlocal body for which the exact solution of a boundary-value problem can be obtained. This is a one-dimentional, semi-infinite body, with the interaction force between material particles x and y given by the function

$$\phi(x,y) = ae^{-\alpha|x-y|} . \qquad (2.1)$$

The equilibrium equation then has the form

$$\int_0^\infty ae^{-\alpha|x-y|}[u(x) - u(y)]dy = f(x). \qquad (2.2)$$

Here f(x) is the density of external force, equilibrated at every point x by the resultant of interactions with all the other particles of the body. The parameter α^{-1} can be interpreted as the effective range of these interactions. As proved later, the constant a is proportional to the elastic modulus E (i.e. Young's modulus for longitudinal, and shear modulus for transverse strain):

$$a = \frac{\alpha^3}{2} E . \qquad (2.3)$$

Let

$$f(x) = F\beta e^{-\beta x} , \qquad (2.4)$$

which is a distribution of force with a variable parameter β and a constant force F. The effective range of the force decreases with increasing β, so that we can interpret the limit of f for $\beta\to\infty$ as a force concentrated on the surface of the body (at the point x = 0). For the sake of convenience we assume E=1, and F=-1 (a tensile force), so that the equation of equilibrium of the body being considered takes the form

$$\frac{\alpha^3}{2}\int_0^\infty e^{-\alpha|x-y|}u(y)dy - \frac{\alpha^2}{2}(2 - e^{-\alpha x})u(x) = $$
$$= \beta e^{-\beta x} . \qquad (2.5)$$

We shall seek a solution of eqn. (2.5) in the form

$$u(x) = \int_0^\infty K(x,y)V(y)dy, \qquad (2.6)$$

where the kernel K(x,y) satisfies the equation

$$\frac{\alpha^3}{2}\int_{-\infty}^{\infty} e^{-\alpha|x-y|}K(y,z)dy - \alpha^2 K(x,z) = \delta(x-z). \qquad (2.7)$$

Hence

$$K(x,y) = |x-y|_+ - \frac{1}{\alpha^2}\delta'(x-y), \qquad (2.8)$$

where the following notation is used

$$|x|_+ = \begin{cases} x , & x \geq 0 \\ 0 , & x < 0 , \end{cases} \qquad (2.9)$$

and $\delta(x)$ denotes Dirac's distribution.

V(x) has to fulfil the equation

$$V(x) + \frac{\alpha^2}{2}e^{-\alpha x}\int_0^\infty K(x,y)V(y)dy = \beta e^{-\beta x} \qquad (2.10)$$

or

$$\left(1 - \frac{1}{2}e^{-\alpha x}\right)V(x) - \frac{\alpha^2}{2}e^{-\alpha x}\int_x^\infty (x-y)V(y)dy =$$
$$= \beta e^{-\beta x} - \frac{\alpha^2}{2}e^{-\alpha x}(Ax - B). \qquad (2.11)$$

Here

$$A = \int_0^\infty V(y)\,dy\ , \qquad B = \int_0^\infty yV(y)\,dy, \tag{2.12}$$

and for a while convergence of both integrals is assumed. The equation (2.11) is of the following form

$$(I + L)V(x) = g(x), \tag{2.13}$$

where I denotes the identity operator and L is defined by

$$LV(x) = -\frac{e^{-\alpha x}}{2}\{V(x) + \alpha^2 \int_x^\infty (x-y)V(y)\,dy\}\ . \tag{2.14}$$

Making use of the perturbation method one obtains the following serie representation of V(x)

$$V(x) = \sum_{n=o}^{\infty} V_n(x) \tag{2.15}$$

the terms of which are connected by the recurrence relations

$$V_o(x) = \beta e^{-\beta x} - \frac{\alpha^2}{2} e^{-\alpha x}(Ax - B), \tag{2.16a}$$

$$V_{n+1}(x) = -LV_n(x)\ . \tag{2.16b}$$

Because of the linearity of the operator I + L, one can write

$$V(x) = \tilde{V}(x) + \tilde{\tilde{V}}(x)\ . \tag{2.17}$$

Here

$$\tilde{V}(x) = \sum_{n=0}^{\infty} \tilde{V}_n(x)\ , \qquad \tilde{\tilde{V}} = \sum_{n=0}^{\infty} \tilde{\tilde{V}}_n(x)\ , \tag{2.18}$$

with

$$\tilde{V}_{n+1} = L\tilde{V}_n, \qquad \tilde{V}_o(x) = \beta e^{-\beta x}, \tag{2.19a}$$

and

$$\tilde{\tilde{V}}_{n+1} = L\tilde{\tilde{V}}_n\ , \qquad \tilde{\tilde{V}}_o(x) = -\frac{\alpha}{2} e^{-\alpha x}(Ax - B). \tag{2.19b}$$

Further one obtains for $n \geq 1$

$$\tilde{V}_n(x) = -\frac{\beta}{2^n}\prod_{k=1}^{n}\left[1-\frac{1}{(k-1+\beta/\alpha)^2}\right]e^{-(n\alpha+\beta)x} \tag{2.20}$$

and, because of the relation

$$\prod_{k=0}^{n-1}\left[1-\frac{1}{(k+a)^2}\right] = \frac{\prod_{k=0}^{n-1}(k+a+1)\prod_{k=0}^{n-1}(k+a-1)}{\prod_{k=0}^{n-1}(k+a)^2} =$$

$$= \frac{\Gamma(n-1+a)\Gamma(n+1+a)}{\Gamma(a-1)\Gamma(a+1)} \cdot \frac{\Gamma^2(a)}{\Gamma^2(n+a)} = \frac{a}{a+1} \cdot \frac{n+a+1}{n+a}, \tag{2.21}$$

one has

$$\tilde{\tilde{V}}_n = \alpha(\varepsilon-1)2^{\varepsilon}\left(1+\frac{1}{n-1+\varepsilon}\right)z^{n+\varepsilon} \tag{2.22}$$

,

where the following notation has been introduced

$$z = \frac{1}{2}e^{-\alpha x} \qquad \varepsilon=\beta/\alpha . \tag{2.23}$$

Similarly

$$\tilde{\tilde{V}}_n = A\alpha\left(1+\frac{1}{n}\right)z^{n+1} . \tag{2.24}$$

Let us introduce, for $z\in(0,\frac{1}{2}\rangle$ and $\varepsilon>0$, a function defined by the relation

$$F(z,\varepsilon) = \int_0^z \frac{y^3}{1-y}dy . \tag{2.25}$$

This is the incomplete Euler beta function $B_z(p,q)$ with $p=\varepsilon+1$, $q=-1$. This function satisfies the following functional equation

$$F(z,\varepsilon-1) - F(z,\varepsilon) = \frac{z^{\varepsilon}}{\varepsilon} , \tag{2.26}$$

and the assymptotic conditions

$$\lim_{\varepsilon\to\infty} F(z,\varepsilon) = 0 = \lim_{z\to 0_+} F(z,\varepsilon) . \tag{2.27}$$

Moreover, the derivative of F, $F'(z,\varepsilon) = \frac{dF}{dz}$, fulfills the equation

$$zF'(z,\varepsilon) = F'(z,\varepsilon+1) . \tag{2.28}$$

The function $\tilde{V}(x)$, with the notation (2.23), is expressed by $F(z,\varepsilon)$ as follows

$$\tilde{V}(x) = \alpha 2^{\varepsilon}\{\varepsilon z^{\varepsilon} + (\varepsilon-1)z[\frac{z^{\varepsilon}}{1-z} + F(z,\varepsilon-1)]\} \tag{2.29}$$

On the other hand, for $\tilde{\tilde{V}}(x)$ we have the expression

$$\tilde{\tilde{V}}(x) = -\alpha\{(A\ln 2z - \alpha B)z + Az[\ln(1-z) - \frac{z}{1-z}]\} . \tag{2.30}$$

We can determine the coefficients A and B, by substituting $\tilde{V} + \tilde{\tilde{V}}$ from (2.29) and (2.30) into (2.12) and then solving these two equations for A and B. As a result we obtain

$$A = 1 , \tag{2.31a}$$

$$B = -\frac{1}{\alpha}\{\ln 2 - \frac{1}{\varepsilon} + 2^{\varepsilon}(\varepsilon-1)F(\tfrac{1}{2},\varepsilon)\} . \tag{2.31b}$$

Finally, after substituting the complete expression for V(x) into (2.6) we obtain the solution of the equation of equilibrium (2.5):

$$u(x,\beta) = -\frac{1}{\alpha}\{\ln 2z - \ln(1-z) + \frac{z}{1-z} + (\varepsilon-1)2^{\varepsilon}[\frac{z^{\varepsilon}}{1-z} + F(z,\varepsilon-1)]\} + C, \tag{2.32}$$

where C is an arbitrary constant. This constant gives no new information about the physics of the problem, because one can see directly from the form of the equation, that it is satisfied by an arbitrary constant. Therefore we shall consider the relative displacement with respect to a fixed point x_o of the body. Let $x_o = 0$ (i.e. $z_o = \frac{1}{2}$). Then the displacements of points of the body with respect to its boundary are described by the function

$$W(x,\beta) = -\frac{1}{\alpha}\{\ln\frac{z}{1-z} + \frac{z}{1-z} + (\varepsilon-1)2^{\varepsilon}[\frac{z^{\varepsilon}}{1-z} + F(z,\varepsilon-1) - F(\tfrac{1}{2},\varepsilon-1) - \frac{1}{2^{\varepsilon}-1}] - 1\} \quad (2.33)$$

which vanishes at the boundary.

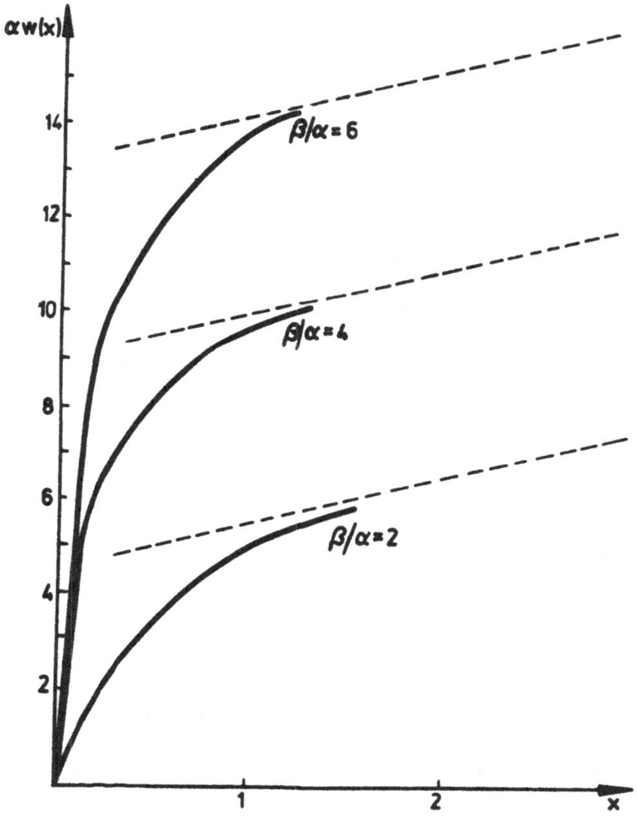

However for any x>0 the value of $W(x,\beta)$ increases when β increases. (Graphs of the function $W(x,\beta)$ for some values of β are presented in the figure).

We have

$$\lim_{\beta\to\infty} W(x,\beta) = \infty, \quad \forall x>0, \tag{2.34}$$

which means that the relative displacement becomes infinite when the force f is concentrated at the surface. Any thin boundary layer responds to the surface force with infinite deformation. In this example, the body cannot, in any realistic way, transfer the forces concentrated at the boundary.

The above example shows, that even the simplest nonlocal bodies can behave in an unusual manner. Bearing this in mind, we shall consider the simplest form of boundary, i.e. a plane. This leads us to examining a layer or a half-space.

Description of the types of nonlocal bodies considered

Expressions for elastic energy We shall now describe the categories of nonlocal bodies mentioned above. First of all, we shall introduce the volume-integral model, which was studied by Kunin[8]. It is a model analogous to the discrete model of the crystal lattice. Interactions between particles are described by a tensor field $\underline{\Phi}(\underline{x},\underline{y})$ (analogous to the crystal dynamical matrix) in such a manner, that the elastic energy of a body Ω is given by the following formula

$$E = -\frac{1}{2}\int_{\Omega}\int_{\Omega}\Phi_{ij}(\underline{x},\underline{y})u_i(\underline{x})u_j(\underline{y})dv(\underline{x})dv(\underline{y}) + \frac{1}{2}\int_{\Omega}A_{ij}(\underline{x})u_i(\underline{x})u_j(\underline{x})dv(\underline{x}), \tag{3.1}$$

where

$$A_{ij}(\underline{x}) = \int_{\Omega}\Phi_{ij}(\underline{x},\underline{y})dv(\underline{y}) \tag{3.2}$$

This form of the functional of energy is implied by the translation invariance.

The next class contains surface-volume integral models, examined by Datta and Kröner[9] and Kröner[10]. For such models, the expression for the elastic energy has the form

$$\begin{aligned} E = &-\frac{1}{2}\int_{\Omega}\int_{\Omega}\Phi_{ij}(\underline{x},\underline{y})u_i(\underline{x})u_j(\underline{y})dv(\underline{x})dv(\underline{y}) + \\ &+ \int_{\Omega}\int_{\partial\Omega}\Psi_{ij}(\underline{x},\underline{y})u_i(\underline{x})u_j(\underline{y})dv(\underline{x})ds(\underline{y}) + \\ &+ \frac{1}{2}\int_{\partial\Omega}\int_{\partial\Omega}\chi_{ij}(\underline{x},\underline{y})u_i(\underline{x})u_j(\underline{y})ds(\underline{x})ds(\underline{y}) + \\ &+ \frac{1}{2}\int_{\Omega}A_{ij}(\underline{x})u_i(\underline{x})u_j(\underline{x})dv(\underline{x}) + \\ &+ \frac{1}{2}\int_{\partial\Omega}B_{ij}(\underline{x})u_i(\underline{x})u_j(\underline{x})ds(\underline{x}), \end{aligned} \tag{3.3}$$

with

$$A_{ij}(\underline{x}) = \int_{\Omega} \Phi_{ij}(\underline{x},\underline{y})\ V(\underline{y}) + \int_{\partial\Omega} \Psi_{ij}(\underline{x},\underline{y}) ds(\underline{y}), \tag{3.4}$$

and

$$B_{ij}(\underline{x}) = \int_{\Omega} \Phi_{ij}(\underline{x},\underline{y}) dv(\underline{y}) + \int_{\partial\Omega} \chi_{ij}(\underline{x},\underline{y}) ds(\underline{y}) \tag{3.5}$$

The symbol $\partial\Omega$ denotes the boundary of Ω. This model differs from the first one by the explicit introduction of the boundary. Besides interactions between interior particles of the body, which are expressed by $\underline{\Phi}(\underline{x},\underline{y})$, one has here interactions between boundary and interior particles, and between boundary particles. They are denoted by $\underline{\Phi}(\underline{x},\underline{y})$ and $\chi(\underline{x},\underline{y})$ respectively.

Finally, we shall write an expression for elastic energy within the integro-differential model:

$$E = \frac{1}{2}\int_{\Omega}\int_{\Omega} \Psi_{ijkl}(\underline{x},\underline{y}) u_{i,k}(\underline{x}) u_{j,l}(\underline{y}) dv(\underline{x}) dv(\underline{y}) + \\ + \frac{1}{2}\int_{\Omega} C_{jkl}(\underline{x}) u_{i,k}(\underline{x}) u_{i,l}(\underline{y}) dv(\underline{x}) \tag{3.6}$$

Here the second term is local and dependent from the first one. A comma denotes partial derivation with respect to space variables. For this model, the interactions between particles depend on the derivatives of the displacement field and not, as the previous models on the displacement directly. Media of such a type was considered by Edelen[11] and Eringen[12].

Basic assumptions on interactions (A) All the interactions decrease sufficiently fast when the distance between interacting particles increases. In particular for all kernels $\underline{K}(\underline{x},\underline{y})$

$$\lim_{|\underline{x}-\underline{y}|\to\infty} \underline{K}(\underline{x},\underline{y}) = 0 \tag{3.7}$$

(B) The body Ω is homogeneous at large distances from the boundary: when particles $\underline{x}$ and $\underline{y}$ are far from the boundary, then $\underline{K}(\underline{x},\underline{y})$ tends to the corresponding function $\underline{K}^a(\underline{x}-\underline{y})$ characterising the unbounded medium. One can say more precisely, that

$$\underline{K}(\underline{x},\underline{y}) = \underline{K}^b(\underline{x},\underline{y}) + \underline{K}^a(\underline{x},\underline{y}), \tag{3.8}$$

where the function $\underline{K}^b(\underline{x},\underline{y})$ decreases with increasing distance from the boundary:

$$(|\underline{x}-\underline{x}_o| + |\underline{y}-\underline{y}_o|\to\infty) \to (\underline{K}^b(\underline{x},\underline{y})\to 0)$$

$$\underline{x}_o,\underline{y}_o \in \partial\Omega,\ \underline{x},\underline{y} \in \Omega. \tag{3.9}$$

(C) In the long wave limit (and far from the boundary), the medium is a classical elastic medium (the local one).

With this assumption one can define the nonlocal elastic modulae.

(D) All kernels are continuous and bounded.

(E) The body is elastic. This implies the following symmetries of the kernels

$$\begin{aligned}\Phi_{ij}(\underline{x},\underline{y}) &= \Phi_{ji}(\underline{y},\underline{x})\\ \chi_{ij}(\underline{x},\underline{y}) &= \chi_{ji}(\underline{y},\underline{x})\\ \Psi_{ijkl}(\underline{x},\underline{y}) &= \Psi_{jilk}(\underline{y},\underline{x})\end{aligned} \tag{3.10}$$

(F) Media are stable, which means that they resist all deformations. The energy of a body is non-negative, and equals zero for rigid displacements only.

Equations of statics By application of the minimum energy principle and by using the Hermitean property of the kernels (3.10), one obtains static equations for the body Ω. Namely one has for VIM:

$$\int_{\Omega}\Phi_{ij}(\underline{x},\underline{y})[u_j(\underline{x}) - u_j(\underline{y})]dv(\underline{y}) = f_i(\underline{x}), \tag{3.11}$$

where $f(\underline{x})$ is an external force density which is, for every $\underline{x}$ equilibrated by interactions of the particle $\underline{x}$ with all the other particles of the body.

For VSIM one obtains, when $\underline{x} \in \Omega$,

$$\int_{\Omega} \Phi_{ij}(\underline{x},\underline{y})[u_i(\underline{x}) - u_j(\underline{y})]dv(\underline{y}) + \tag{3.12a}$$

$$+ \int_{\partial\Omega} \Psi_{ij}(\underline{x},\underline{y})[u_j(\underline{x}) - u_j(\underline{y})]ds(\underline{y}) = f_i(\underline{x})$$

and when $\underline{x} \in \partial\Omega$

$$\int_{\Omega} \Psi_{ij}(\underline{x},\underline{y})[u_j(\underline{x}) - u_j(\underline{y})]dv(\underline{y}) + \tag{3.12b}$$

$$+ \int_{\partial\Omega} \chi_{ij}(\underline{x},\underline{y})[u_j(\underline{x}) - u_j(\underline{y})]ds(\underline{y}) = p_i(\underline{x})$$

where $\underline{p}(\underline{x})$ is the density of the external surface force.

Finally, for IDM, when $\underline{x} \in \Omega - \partial\Omega$

$$\frac{\partial}{\partial x_k}\{\int_{\Omega} \Psi_{ijkl}(\underline{x},\underline{y})u_{j,l}(\underline{y})dv(\underline{y}) + \tag{3.13a}$$

$$+ C_{ijkl}(\underline{x})u_{j,l}(\underline{x})\} = f_i(\underline{x})$$

and, for $\underline{x} \in \partial\Omega$

$$\{\int_{\Omega} \Psi_{ijkl}(\underline{x},\underline{y})u_{j,l}(\underline{y})dv(\underline{Y}) + \tag{3.13b}$$

$$+ C_{ijkl}(\underline{x})u_{j,l}(\underline{x})\}n_k(\underline{x}) = p_i(\underline{x})$$

where $\underline{n}(\underline{x})$ is the vector normal to $\partial\Omega$ at the point $\underline{x}$.

Transition to the one-dimensional case Let us assume, that all components of displacement fields (and of forces) are functions dependent on only one variable x_1, the coordinate normal to the boundary planes of the layer under consideration.

$$u_i = u_i(x_1), \qquad a \leqslant x_1 \leqslant b \tag{3.14}$$

Moreover, let us assume, that the form of the interactions assures invariance of static equations with respect to:

(a) translations parallel to the surfaces,

(b) rotations around the axis orthogonal to these planes, and,

(c) reflections with respect to every plane containing this axis.

The above assumptions imply, that the kernel $\underline{\Phi}(\underline{x},\underline{y})$ has a representation

$$\underline{\Phi} = \begin{pmatrix} \tilde{a} , & y_2 c , & y_3 d \\ y_2 c , & \tilde{b} , & y_2 y_3 e \\ y_3 d , & y_2 y_3 e , & \tilde{b} \end{pmatrix} \tag{3.15}$$

where $\tilde{a}$, $\tilde{b}$, c, d, e are functions dependent on x_1, y_1 and on the invariant $\rho \overset{df}{=} \sqrt{(x_2 - y_2)^2 + (x_3 - y_3)^2}$.

Because of the parity of the functions c, d and e with respect to variables y_2 and y_3, one obtains:

$$\int_{-\infty}^{\infty}\int_{-\infty}^{\infty} y_2 c(x_1,y_1,\rho)dy_2dy_3 = 0,$$

$$\int_{-\infty}^{\infty}\int_{-\infty}^{\infty} y_2 c(x_1,y_1,\rho)dy_2dy_3 = 0, \tag{3.16}$$

$$\int_{-\infty}^{\infty}\int_{-\infty}^{\infty} y_2 y_3 e(x_1,y_1,\rho)dy_2dy_3 = 0.$$

Let us assume, that a and b decrease at infinity at least as quickly as $\rho^{-(3+\varepsilon)}$, $\varepsilon>0$. Then one can introduce functions $a(x_1,y_1)$ and $b(x_1,y_1)$ defined by the formulae:

$$a(x_1,y_1) = \int_{-\infty}^{\infty}\int_{-\infty}^{\infty}\tilde{a}(x_1,y_1,\rho)dy_2dy_3,$$
$$b(x_1,y_1) = \int_{-\infty}^{\infty}\int_{-\infty}^{\infty}\tilde{b}(x_1,y_1,\rho)dy_2dy_3 \tag{3.17}$$

Taking into account equations (3.15) and (3.16) one can transform equations (3.11) into a separated system of equations for components of displacements:

$$\int_a^b a(x,y)[u_1(x) - u_1(y)]dy = f_1(x),$$
$$\int_a^b b(x,y)[u_2(x) - u_2(y)]dy = f_2(x), \tag{3.18}$$
$$\int_a^b b(x,y)[u_3(x) - u_3(y)]dy = f_3(x)$$

The first equation describes tension or compression of the body by forces orthogonal to the surfaces. The remaining two equations concern the shear deformation in two directions. One sees immediately, that all these equations are of the same form. Hence for the volume-integral model it is sufficient to investigate the following one-dimensional equation:

$$\int_a^b \phi(x,y)[u(x) - u(y)]dy = f(x) \tag{3.19}$$

where

$$\phi(x,y) = \phi(y,x) \tag{3.19a}$$

Let us notice, that the example presented in chapter I describes a particular case of such a body.

In a similar way eqs. (3.12) can be transformed into three triples of an identical form (every triple corresponds to one component of displacements field). As a result one obtains:

$$\int_a^b \phi(x,y)[u(x) - u(y)]dy + \psi(x,a)[u(x) - u(a)] +$$

$$+ \psi(x,b)[u(x) - u(b)] = f(x) \tag{3.20a}$$

$$\int_a^b \psi(x,a)[u(a) - u(x)]dx + \zeta(a,b)[u(a) - u(b)] = p\ , \tag{3.20b}$$

$$\int_a^b \psi(x,b)[u(b) - u(x)]dx - \zeta(a,b)[u(a) - u(b)] = q\ , \tag{3.20c}$$

where the kernels equal the relevant components of tensor kernels $\underline{\Phi}, \underline{\Psi}$ and $\underline{X}$ integrated with respect to the variables y_2 and y_3. For example, for the kernels corresponding to the first component of $\underline{u}$ one has the following formulae:

$$\phi(x,y) = \int_{-\infty}^{\infty}\int_{-\infty}^{\infty} \Phi_{11}(x_1, y_1, \rho)dy_2 dy_3, \tag{3.21a}$$

$$\psi(x,a) = \int_{-\infty}^{\infty}\int_{-\infty}^{\infty} \Psi_{11}(x_1, a, \rho)dy_2 dy_3, \tag{3.21b}$$

$$\zeta(a,b) = \int_{-\infty}^{\infty}\int_{-\infty}^{\infty} \chi_{11}(a,b,\rho)dy_2dy_3 \qquad (3.21c)$$

It is assumed, that the integrals involved are convergent. Then the right hand sides of (3.20) are equal to:

$$p = p_1(a), \; q = p_1(b) \qquad (3.21d)$$

We assume the symmetry of the body with respect to the middle plane parallel to the surfaces:

$$\psi(a+x,a) = \psi(b-x,b), \quad x \in \langle a,b\rangle \qquad (3.22)$$

Besides equations (3.20) we have the equilibrium condition:

$$\int_a^b f(x)dx + p + q = 0, \qquad (3.23)$$

which permits us to eliminate one of these equations.

The above procedure applied for a third time leads to separation of the static equations for the integro-differential model. As a result one obtains three triples of the form:

$$-\frac{\partial}{\partial x}\{\int_a^b \Theta(x,y)u'(y)dy - C(x)u'(x)\} = f(x), \qquad (3.24a)$$

$$-\int_a^b \Theta(a,x)u'(x)dx - C(a)u'(a) = p, \qquad (3.24b)$$

$$\int_a^b \Theta(b,x)u'(x)dx - C(b)u'(b) = q \qquad (3.24c)$$

where the dash denotes differentiation and, for example

for the first component of $\underline{u}$, one has

$$\Theta(x,y) = \int_{-\infty}^{\infty}\int_{-\infty}^{\infty}\Psi_{1111}dy_2dy_3, \tag{3.25a}$$

$$C(x) = C_{1111}(x), \tag{3.25b}$$

$$p = p_1(a), \quad q = p_1(b) \tag{3.25c}$$

The kernel Θ is symmetrical

$$\Theta(x,y) = \Theta(y,x) \tag{3.26}$$

IV. INFINITE MEDIA

This chapter is based on the papers[13-14]

Let us first notice, that for infinite bodies there is no difference between media described by VIM and SVIM. This conclusion falls precisely under our assumptions on behaviour of the kernels far from the boundary. Both integral media have then an equilibrium equation of the form

$$\int_{-\infty}^{\infty} K(\xi)\,[u(x) - u(x+\xi)]\,d\xi = f(x) \qquad (4.1)$$

The assumption of homogeneity for IDM implies, that

$$C(x) = \text{const} = D \qquad (4.2)$$

and then the equilibrium equation has the form

$$-\frac{\partial}{\partial x}\{\int_{-\infty}^{\infty} N(\xi)\,u'(\xi+x)\,d\xi + Du'(x)\} = f(x) \qquad (4.3)$$

1. Nonlocal elastic modulae

The Fourier representation of the equilibrium equation can be written

$$\Lambda(k)\hat{u}(k) = \hat{f}(k) \qquad (4.4)$$

where $\hat{u}$ and $\hat{f}$ denote the Fourier transforms of u and f respectively, and Λ is a continuous function of the wave number k. From the assumptions of symmetry and elasticity it follows, that $\Lambda(k)$ is real and even. The stability assumption requires that

$$\Lambda(k) > 0 \text{ for } k \neq 0 \tag{4.5}$$

and the correspondence with classical elasticity in the long-wave limit requires that

$$\Lambda(k) = Ek^2 + \sigma(k^2) \text{ for } k \to 0 \quad , \tag{4.6}$$

where E is the classical elastic modulus and $\sigma(k^2)$ denotes a quantity of higher order with respect to k^2.

For our integral models the function $\Lambda(k)$ has the form

$$\Lambda(k) = \hat{K}(0) - \hat{K}(k) \tag{4.7}$$

If one assumes that the kernel $K(\xi)$ has summable moments up to second order,

$$\int_0^\infty (1+\xi)^2 |K(\xi)| d\xi < \infty, \tag{4.8}$$

then $\Lambda(k)$ is a function belonging to C^2, which assures the existence of the expansion (4.6). Hence one obtains

$$\hat{K}(k) < \hat{K}(0) \text{ for } f \neq 0, \tag{4.9}$$

and the following expression for the nonlocal Young's modulus:

$$E = -\frac{1}{2}K''(0) = \int_0^\infty \xi^2 K(\xi) d\xi \quad . \tag{4.10}$$

In the case of the integro-differential model one assumed that the kernel is summable

$$\int_0^\infty |N(\xi)| d\xi < \infty \quad . \tag{4.11}$$

The function $\Lambda(k)$ for this model is given by

$$\Lambda(k) = k^2(\hat{N}(k) + D) \tag{4.12}$$

and, taking into account (4.5) and (4.6) one obtains the inequality

$$\hat{N}(k) + D > 0 \text{ for every } k, \tag{4.13}$$

and the following expression for the Young's modulus

$$E = D + \hat{N}(0) \tag{4.14}$$

or

$$E = D + \int_{-\infty}^{\infty} N(\xi)d\xi \quad . \tag{4.15}$$

2. Homogeneous equation

Relations (4.5) and (4.6) lead to a simple but important conclusion concerning the uniqueness of solutions in the nonlocal theory of elasticity. In the present one-dimensional case we have

Theorem 3.1

If $f=0$, then

$$u(x) = ax + b \quad . \tag{4.16}$$

Theorem 3.2

If for given f a solution u exists in the class of temperate distributions, then it is unique modulo the class (4.16).

This result is physically plausible. The term b in (4.16) represents a rigid displacement. (Notice, that in the three-dimensional case one also has to take into account rigid rotations). To explain the presence of the uniform deformation ax let us note, that f=0 does not mean that there are no forces acting on the medium. It only means, that there are no forces acting at finite x's. The term ax can therefore be interpreted as due to certain forces applied at infinity.

3. The existence of fundamental solutions

For the sequel it is important to distinguish between the degenerate and nondegenerate cases in our models. Namely, we shall consider the cases K(0)=0 for the integral models and D=0 for the integro-differential one as degenerate. On the contrary, we have the inequalities

$$\hat{K}(0) > 0 \quad \text{and} \quad D > 0 \tag{4.17}$$

The impossibility of negative values in (4.17) follows from (4.9) and (4.13) by the relations

$$\hat{K}(k) \to 0, \; \hat{N}(k) \to 0 \quad \text{for } k \to \infty \tag{4.18}$$

A fundamental solution to equation (4.1) or(4.3) is, by definition, a solution corresponding to Dirac's delta:

$$f(x) = \delta(x) \tag{4.19}$$

If it exists in the class of temperate distributions, it is, according to the Theorem 3.2 unique modulo the class (4.16).

We have

Theorem 3.3

In the nondegenerate case the fundamental solutions for unbounded media exists in both nonlocal models.

Proof

Because of (4.4) and (4.5), for the existence of a fundamental solution in a class of temperate distributions it is necessary and sufficient, that the function $\Lambda^{-1}(k)$ be of temperate growth at infinity. This condition is satisfied in the nondegenerate case as a consequence of relations (4.18). Q.E.D.

In the degenerate case the fundamental solutions do not always exist, as Barnett's example shows[15]. Generally, it follows from the theorems on the singular order of operators that the degenerate models with infinitely differentiable kernels have no fundamental solutions in the class of temperate distributions.

In general, whenever $\Lambda^{-1}(k)$ is of temperate growth at infinity, the formula

$$\hat{G}(k) = \Lambda^{-1}(k) \tag{4.20}$$

with an appropriate regularization procedure for $k \to 0$ defines the Fourier transform of a fundamental solution G(x) which is always a temperate distribution.

3. An auxiliary equation

For the sake of brevity we introduce the following mode of expression. Let ϕ be a function of a real variable. We shall say that ϕ is B if it is bounded, C if it is continuous, L_1 if it is summable, L_2 if its square is summable. If P is a certain property of ϕ, then the expression "ϕ is P_s^q" means that the property P is shared by all the functions

$$x^r \phi^{(p)}(x), \quad 0 \leqslant r \leqslant s, \quad 0 \leqslant p \leqslant q \tag{4.21}$$

where p, q, r and s are integers. The conjunction of two properties P and R is denoted by PR. The expression "ϕ is $\dot{P}$" means that ϕ has the property P and also tends to zero at infinity.

Consider first the auxiliary equation

$$\int_{-\infty}^{\infty} Q(\xi)\Delta(x+\xi)d\xi + \Delta(\xi) = \delta(x) \tag{4.22}$$

under the assumption that Q(k) is continuous and

$$\hat{Q}(k) + 1 > 0, \quad \hat{Q}(\pm\infty) = 0 \tag{4.23}$$

Using the identity:

$$(1 + Q)^{-1} = 1 - Q(1 + Q)^{-1} \qquad (4.24)$$

one can represent the solution (4.22) in the form

$$\Delta(x) = \delta(x) - \Gamma(x) \qquad (4.25)$$

where

$$\hat{\Gamma}(k) = \hat{Q}(k)(1 + \hat{Q}(k))^{-1} \qquad (4.26)$$

Now for the function

$$\Gamma_1 = Q - \Gamma \qquad (4.27)$$

we have the following Lemmas.

Lemma 3.1

If $Q(\xi)$ is $(BL_1)^q_s$, then $\hat{\Gamma}_1(k)$ is $(\dot{C}L_1)^s_{2q}$.

Proof

(a) The Fourier transform of Γ_1 equals

$$\hat{\Gamma}_1(k) = \hat{Q}^2(k)[1 + \hat{Q}(k)]^{-1} \qquad (4.28)$$

(b) Let q=s=0. Since BL_1 implies L_2, then by the Riemann-Lebesque Lemma and the Parceval identity, the transform $\hat{Q}$ is $\dot{C}L_2$ and, in consequence, Γ_1 is CL_1.

(c) Let q be arbitrary, s=0. By a similar argument $\hat{Q}$ is $(\dot{C}L_2)_q$ and Γ_1 is $(\dot{C}L_1)_{2q}$.

(d) Let q, s be arbitrary. The induction with respect to s shows that $\hat{Q}$ is $(\dot{C}L_2)^s_q$. Now notice, that the product of two functions which are $(\dot{C}L_2)_q$ and $(\dot{C}L_2)_p$ respectively, is $(\dot{C}L_2)_{p+q}$. In consequence, all the derivatives of Γ_1 up to the order s are $(\dot{C}L_1)_{2q}$ and, in conclusion, Γ_1 is $(\dot{C}L_1)^s_{2q}$. Q.E.D.

Lemma 3.2

For arbitrary ϕ, if $\hat{\phi}$ is $(\dot{C}L_1)^s_p$, then ϕ is $(\dot{C}L_2)^p_s$

Proof

The property $(L_1)^s_p$ of implies $\dot{C}P_s$ of $\hat{\phi}$. Since CL_1 implies L_2, the property $(L_2)^s_p$ is Fourier-transformed into $(L_2)^p_s$, and the Lemma follows. Q.E.D.

For the function $\Gamma(x)$ in (4.25) we have the following

Theorem 3.3

If $Q(\xi)$ is $(BCL_1)^q_s$, then $\Gamma(\xi)$ is $(BCL_2)^q_s$. If $Q(\xi)$ is $(\dot{C}L_1)^q_s$, then $\Gamma(\xi)$ is $(\dot{C}L_2)^q_s$.

Proof

This follows simply from the above lemmas and relation (4.27).

4. The smoothness and asymptotic behaviour of the fundamental solutions

Let us now consider the integral medium. We introduce the notation

$$A = \hat{K}(0) = \int_{-\infty}^{\infty} K(\xi)d\xi \quad . \tag{4.29}$$

According to the definitions of section 2, the kernel $K(\xi)$ is $(CL_1)_2$. If besides that it tends to zero at infinity sufficiently fast, the following theorem can be effectively applied.

Theorem 3.4

Let the kernel $K(\xi)$ be $(BCL_1)'^{q}_{s}$ with some q and s. If

$A \neq 0$

(i) the fundamental solution can be represented in the form

$$G(x) = \frac{1}{A}\delta(x) + \frac{1}{2E}|x| - H(x), \tag{4.30}$$

where A is given by (4.29) and E is the elastic modulus (4.10).

(ii) if $q \geqslant 2$, $s \geqslant 2$, then $H''(x)$ is $(BCL_2)^{q-2}_{s-2}$

(iii) if $q \geqslant 0$, $s \geqslant 5$, then $H(x)$ is $(BC)^{q}_{s-q}$.

Proof

Let $$\hat{Q}(k) = -\frac{1}{A}\hat{K}(k) + \frac{1}{Ek^2}(A - \hat{K}(k)) \tag{4.31}$$

Then, identically,

$$[A - \hat{K}(k)]^{-1} = (\frac{1}{A} + \frac{1}{Ek^2})[1 + \hat{Q}(k)]^{-1}; \qquad (4.32)$$

or

$$\hat{G}(k) = (\frac{1}{A} + \frac{1}{Ek^2})\hat{\Delta}(k) \qquad (4.33)$$

which proves the existence of representation (4.30) with H given by the following expression:

$$H(x) = \frac{1}{A}\Gamma(x) + \frac{1}{2E}\int_0^x (x-y)\Gamma(y)dy + C, \qquad (4.34)$$

where C is an arbitrary constant. If follows from (4.31), that if $\hat{K}(k)$ is $(\dot{C}L_2)^s_q$, then $\hat{Q}(k)$ is $(\dot{C}L_2)^{s-2}_q$ and, in consequence, $\Gamma(x)$ is $(BCL_2)^q_{s-2}$. If $q \geqslant 2$, assertion (ii) follows from (4.34). Further, let us note that according to (4.31) we have $\hat{Q}(0)=0$ and, by (4.26), also $\Gamma(0)=0$. Then, since $\Gamma(y)$ is even, we have

$$\int_0^\infty \Gamma(y)dy = 0, \quad \int_{-\infty}^\infty y\Gamma(y)dy = 0 \quad , \qquad (4.35)$$

the integrals being well convergent for $s \geqslant 5$. Taking

$$C = \int_0^\infty y\Gamma(y)dy \qquad (4.36)$$

we can rewrite (4.34) as

$$H(x) = \frac{1}{A}\Gamma(x) + \frac{1}{2E}\int_{-\infty}^x (x-y)\Gamma(y)dy = \qquad (4.37)$$

$$= \frac{1}{A}\Gamma(x) - \frac{1}{2E}\int_x^\infty (x-y)\Gamma(y)dy \quad .$$

If $s \geq 5$, then the integrals in (4.36) are $(BC)^{q+2}_{s-4}$ and assertion (iii) follows. Q.E.D.

In the models of integro-differential type, by passing to the new dependent variables $\tau(x)$ and $\varepsilon(x)$, where

$$\tau(x) = \int_{-\infty}^{\infty} N(\xi)u'(x+\xi)d\xi + Du'(\xi) \qquad (4.38a)$$

and

$$\varepsilon(x) = u'(x) \qquad (4.38b)$$

the governing equation (4.3) can be converted into

$$\int_{-\infty}^{\infty} N(\xi)\varepsilon(x+\xi)d\xi + D\varepsilon(x) = \tau(x) \quad . \qquad (4.39)$$

If $D \neq 0$, this equation is of the auxiliary type (4.22) and with

$$Q(\xi) = \frac{1}{D}N(\xi) \qquad (4.40)$$

we obtain

$$\varepsilon(x) = \frac{1}{D}\tau(x) - \frac{1}{D}\int_{-\infty}^{\infty}\Gamma(\xi)\tau(x+\xi)d\xi \quad . \qquad (4.41)$$

Theorem 3.1 is directly applicable here.

For the fundamental solution $G(x)$ in the non-degenerate IDM we have

Theorem 3.5

If the kernel $N(\xi)$ is summable and $D \neq 0$, then the fundamental solution is continuous.

Proof

For the Fourier transform of the fundamental solution in the IDM we have the expression

$$\hat{G}(k) = \frac{1}{k^2}(N(k) + D)^{-1} \tag{4.42}$$

and by the Paley-Wiener theorem and the Riemann-Lebesque lemma the conclusion of the theorem follows.

Theorem 3.6

If the kernel $N(\xi)$ is $(BCL_1)^q_s$ with some $q \geq 0$ and $s \geq 0$, and if $D \neq 0$, then there exists a fundamental solution of the form

$$G(x) = \frac{1}{2D}|x| - H(x), \tag{4.43}$$

where $H''(x)$ is $(BCL_2)^q_s$ and $H(x)$ is C^{q+s}.

Proof

With (4.40) we have

$$H''(x) = \frac{\Gamma(x)}{D} \tag{4.44}$$

and Theorem 3.1 is applicable.

Theorem 3.7

Under the assumptions of Theorem 3.6, the fundamental solution (4.43) can also be represented in the form

$$G(x) = \frac{1}{2E}|x| - H^{as}(x), \tag{4.45}$$

where

(i) if $s \geq 2$, then $H^{as\prime}(x)$ is $(BC)_{s-1}$

(ii) If $s \geq 3$, then $H^{as}(x)$ is $(BC)_{s-2}$.

Proof

(ii) Let

$$H^{as}(x) = \int_{|x|}^{\infty} (|x| - y)H''(y)dy \tag{4.46}$$

and

$$H(x) = x\int_0^x H''(y)dy + \int_x^{\infty} yH''(y)dy. \tag{4.47}$$

Then

$$H(x) = a|x| + H^{as}(x), \tag{4.48}$$

where

$$a = \int_0^{\infty} H''(y)dy = \frac{1}{2D}\hat{\Gamma}(0) = \frac{E - D}{2DE}. \tag{4.49}$$

By substituting (4.48) into (4.43) we obtain representation (4.45) and assertion (ii) follows from (4.46) and Theorem 3.6.

(i) By a similar argument for the derivative

$$H^{as\prime}(x) = -\int_{|x|}^{\infty} H''(y)dy \tag{4.50}$$

the assertion (i) is proved. Q.E.D.

5. The existence of continuous solutions

The theorems concerning the smoothness and the asymptotic behaviour of the fundamental solutions established in the previous sections can be directly applied to investigating the corresponding properties of other solutions and to solving some existential questions, such as the problem of the existence of continuous solutions.

Theorem 3.8

Let the kernel $K(\xi)$ of the non-degenerate integral model be $(BCL_1)_3$. If, moreover, the force density is continuous and satisfies the following condition

$$\int_{-\infty}^{\infty}(1 + |x|)|f(x)|dx < \infty, \tag{4.51}$$

then

(i) a solution $u(x)$ exists,

(ii) any solution $u(x)$ is continuous, and

(iii) for any solution $u(x)$ there exists a constant C such that

$$|u(x)| \leq C(1 + |x|) \tag{4.52}$$

Proof

If K is $(BCL_1)_3$, then Γ is $(BC)_1$, and

$$|H(x)| \leq C_1(1 + |x|) \tag{4.53}$$

for some C_1. In consequence of (4.51) and (4.52) the following formula

$$u(x) = \frac{1}{A}f(x) + \frac{1}{2E}\int_0^x (x-y)f(x)\,dy - \int_{-\infty}^{\infty} H(x-y)f(y)\,dy \tag{4.54}$$

defines u which is continuous and satisfies (4.52) with a certain constant C. On substituting (4.54) to the governing equation (4.1), one verifies that the integrals involved are absolutely convergent, and the equation is satisfied. According to Theorem 3.2, any other solution differs from (4.54) by ax + b. Therefore it is also continuous and satisfies (4.52).

Theorem 3.9

Let the kernel $N(\xi)$ of the non-degenerate integro-differential model be BCL_1, and let f(x) be a measurable function, which satisfies (4.51). Then

(i) a solution u(x) which is C^1 exists,

(ii) any solution u(x) is C^1, and

(iii) the derivative u'(x) is bounded.

Proof

If N is BCL_1, then Γ is B_1C, H is C^2 and satisfies (4.53). The following formula

$$u(x) = \frac{1}{2E}\int_0^x (x - y) f(y) dy - \int_{-\infty}^{\infty} H^{as}(x - y) f(y) dy \tag{4.55}$$

defines u which is $C_1 B$ and satisfies the governing equation (4.3) together with the corresponding convergence requirements. Q.E.D.

V. FINITE BODIES

1. Basic definitions

According to the assumptions of the previous chapters we shall examine a finite body in the form of a layer. In the one-dimensional case cosidered here, an interval [a,b] is the geometric model of the body.

Let us introduce some definitions that will be of help in a precise formulation of the question raised in Chapter II concerning the transition from near-surface to surface loads.

Definition 1

A sequence $\{f_n\}$ of body forces on $[0,\infty)$ will be called a VS-sequence if it fulfills the following conditions:

(A) each f_n is monotonic and continuous,

(B) $\int_0^\infty f_n(x)\,dx$ = const. (independent of n),

(C) supp $f_n \subset [0,\varepsilon_n]$, where $\varepsilon_n \xrightarrow[n\to\infty]{} 0$

e.g.

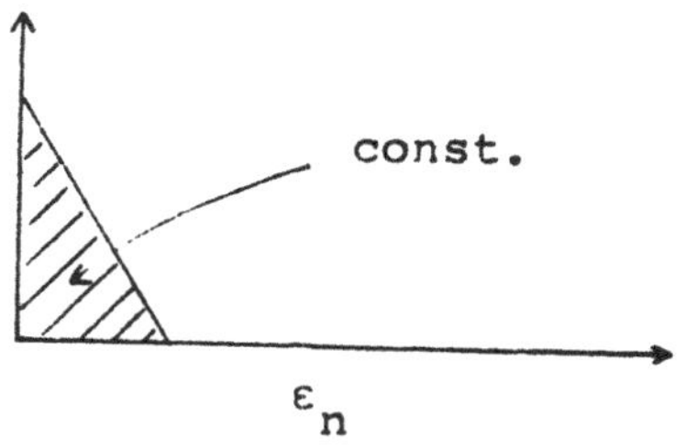

(5.1)

Definition 2

A sequence $\{f_n\}$ of body forces on [a,b] will be called a VS-sequence if it is of the form

$$f_n(x) = \tilde{f}_n(x-a) + \approx{f}_n(b-x) \quad , \qquad (5.2)$$

where $\{\tilde{f}_n\}$ and $\{\approx{f}_n\}$ are VS-sequences on $[0,\infty)$.

Definition 3

Let S be a space of functions. A model will be called VS-compatible in S if for an arbitrary VS-sequence of loads

(A) corresponding solutions u_n belong to S,

(B) the sequence $\{u_n\}$ is convergent and its limit u belongs to S,

(C) this limit equals the solution $\tilde{u}$ for the corresponding surface loads.

In what follows S will be either C - the space of functions continuous on [a,b], or $\mathcal{B}$ - the space of functions continuous on (a,b) and bounded on [a,b]. In the next sections we shall investigate VS-compatibility of the categories of models introduced above.

2. The volume-integral model

We shall consider the non-degenerate case, i.e. the condition

$$A(x) = \int_a^b \phi(x,y)dy \neq 0 \tag{5.3}$$

will be fulfilled. The equilibrium equation can be divided by $\sqrt{A(x)}$ and this operation transforms it into a Fredholm equation of the second kind. Because of stability of the medium, the only solution to this equation with $f(x) = 0$ is $u(x) = \text{constant}$. In consequence, Fredholm's disjunction for this case can be formulated as following

Theorem 4.1

The solution u of equation (3.19) with (5.3) exists in C, for $f \in C$, if and only if the following condition is satisfied

$$\int_a^b f(x)dx = 0 \tag{5.4}$$

This solution is unique up to a constant.

The condition (5.4) has a simple physical interpretation: it means that the resultant of external forces is equal to zero.

Let now a=-b and let us denote by $\overset{s}{C}$ the space of odd functions continuous on the interval [-b,b].

Theorem 4.2

If ϕ is continuous and satisfies (5.3), and if $f\in\overset{s}{C}$, then $\overset{s}{C}$ contains exactly one solution u to the equation·

$$\int_{-b}^{b}\phi(x,y)[u(x) - u(y)]dy = f(x) \quad . \tag{5.5}$$

Proof

If $f\in\overset{s}{C}$, then the condition (5.4) must be satisfied. Hence the Theorem 4.1 assures the existence of a continuous solution u(x) to equation (5.5). Because of linearity of the equation the odd function

$$V(x) = \frac{1}{2}[u(x) - u(-x)] \tag{5.6}$$

is the solution to equation (5.5). It is necessarily unique because after addition of a non-zero constant the function would cease to be odd. Q.E.D.

The above theorem will be of help in proving the following basic theorem

Theorem 4.3

The volume integral non-degenerate model is not VS-compatible in C.

Proof

Let us consider a VS-sequence $\{p_n(x)\}$ on $[0,\infty)$.

Then

$$f_n(x) = p_n(x+b) - p_n(b-x) \qquad (5.7)$$

form an odd VS-sequence on [-b,b]. The limit f of this sequence does not belong to C. Each f_n determines uniquely in C the solution u_n of the static equation. The Fredholm theory assures that this solution can be represented in the form

$$u_n(x) = \frac{1}{A(x)} f_n(x) + v_n(x), \qquad (5.8)$$

where the function $v_n(x)$ is given by the formula

$$v_n(x) = \int_{-b}^{b} \frac{R(x,y)}{[A(x)A(y)]^{1/2}} f_n(y)\,dy \qquad (5.9)$$

$R(x,y)$ denotes here the Fredholm resolvent of the equation (4.3), divided by $\sqrt{A(x)}$.

Because of continuity of the function

$$R(x,y)\,[A(x)A(y)]^{-1/2}$$

the limit v of the sequence v_n belongs to C. In consequence, the limit u of the sequence u_n does not belong to C. Hence, the point (B) of Definition 3 is not satisfied. The model is not VS-compatible in C. Q.E.D.

The above result means that the example discussed at the beginning of our course was not an exception. The relative displacement with respect to the (left) boundary can be written as

$$W_n(x) = \frac{1}{A(x)} p_n(0) + 0(x), \tag{5.10}$$

where $0(x)$ denotes a bounded function on $[-b,b]$. For each $x \in (-b,b)$, if u is sufficiently large, so that the support of p_n is so little, that $p_n(x+b)=0=p_n(b-x)$, $p_n(2b)=0$. From the definition of VS-sequence it follows that

$$\lim_{n\to\infty} p_n(0) = \infty, \tag{5.11}$$

therefore

$$\lim_{n\to\infty} W_n(x) = \infty, \ \forall_x \in (-b,b) \quad . \tag{5.12}$$

The body responds to surface force by an infinite deformation.

3. Surface-volume integral models

Let us note that in the equation of statics of the volume-integral model the surface forces do not appear in an explicit way. The natural question arises whether this is not the cause of the unexpected response of the body to the boundary forces. To explain this we shall consider the models which take into account surface forces. First let us discuss the SVIM. Within this model the medium will be called absolutely stable if it satisfies the following conditions.

(S1) $\phi(x,y)\geq 0,\ \psi(x,y)\geq 0,\ \zeta\geq 0,$

(S2) $\forall x \in [a,b]\quad \exists y \in [a,b]\quad \phi(x,y) \neq 0.$

This means that all the "springs" that model the interactions between particles are stable, i.e. resist deformations. The absolute stability implies the global one.

The SVIM will be called non-degenerate if $\psi(x)\not\equiv 0$.

Let us first consider the degenerate case, i.e. assume that $\psi(x)\equiv 0$. Then the equilibrium equations (3.20) take the form of the following system.

$$\int_a^b \phi(x,y)[u(x) - u(y)]dy = f(x), \tag{5.13a}$$

$$\zeta(a,b)[u(a) - u(b)] = p, \tag{5.13b}$$

$$\zeta(a,b)[u(b) - u(a)] = q. \tag{5.13c}$$

We have for this case

Theorem 4.4

The system (5.13) has a solution in C if the following conditions are satisfied simultaneously.

(a) $\int_a^b \phi(x,y)dy \not\equiv 0,$

(b) $f \in C,$

(c) $\int_a^b f(x)dx = 0,$

(d) $p = -q$,

(e) q, p and f are connected by a relation implied by the following circumstances.

(e1) Given f which fulfils (b) and (c) determines a solution u of the Fredholm equation (5.13a), up to a constant.

(e2) The difference $u(a)-u(b)$ is then unique, therefore p and q are also uniquely determined by (5.13b) and (5.13c).

It is then evident that the system (5.13) in general has no solutions in C.

A similar situation arises in the non-degenerate case.

Theorem 4.5

The system of static equations (3.20), with $\psi(x)\not\equiv 0$ in general has no solutions in C.

Proof

(A) Assume $a=-b$ and consider displacements from $\overset{s}{C}$. We have then

$$p = q,\ f(x) = f(-x), \tag{5.14a}$$

$$\phi(x,y) = \phi(-x,-y),\ \psi(x,b) = \psi(-x,-b) \tag{5.14b}$$

The equilibrium equations take the form

$$\int_{-b}^{b}\phi(x,y)[u(x) - u(y)]dy + \Phi(x)[u(x) - u(b)] =$$

$$= f(x), \qquad (5.15a)$$

$$\frac{1}{2}\int_{-b}^{b}\Phi(x)[u(x) - u(b)]dx = -p, \qquad (5.15b)$$

where the following symbol has been used

$$\Phi(x) = \psi(x,b) + \psi(-x,b). \qquad (5.16)$$

If the system (5.15) has no solutions in $\overset{s}{C}$, then the system (3.20) has no solutions in C, because the symmetrical part of the solution to the system (3.20) would fulfil equations (5.15).

(B) Let us introduce the notation

$$\overset{o}{C} = \{f \in C: f(-b)=0=f(b)\}, \qquad (5.17)$$

and consider the following auxiliary set of equations

$$\int_{-b}^{b}\phi(x,y)[u(x) - u(y)]dy + \Phi(x)u(x) = f(x), \qquad (5.18a)$$

$$\frac{1}{2}\int_{-b}^{b}\Phi(x)u(x)dx = -p. \qquad (5.18b)$$

If a solution of the system (5.15) exists in C, then a solution to the system (5.18) exists in $\overset{o}{C}$.

(C) For $f \in C$, due to the assumed stability of the medium, equation (5.18a) has in C only one solution. This implies that, for a given f, a solution to the system (5.18) exists for at most one value of p.

(D) Let us choose f as follows

$$f(x) \overset{df}{==} \int_{-b}^{b} \phi(x,y)[w(x) - w(y)]dy + \Phi(x)w(x), \tag{5.19}$$

where $w \in C$ and $w(b) \neq 0$.

For such f the only candidate for being a solution to the system (5.18) is the function w(x), but this does not belong to $\overset{\circ}{C}$. Taking into account the point (B), we see that the system (5.15) with f determined by (5.19) has no solutions. Hence the space C contains no solution to equations (3.20). Q.E.D.

One can see that even the point (A) of Definition 3 is not satisfied. Hence

<u>Theorem 4.6</u>

The surface-volume integral model is not VS-compatible in C.

The point (A) of Definition 3 can be, however, fulfilled in the space $\mathcal{B}$. Namely we have

Theorem 4.7

For absolutely stable and nondegenerate SVIM, and for each $f \in C$, p and q fulfilling the equilibrium condition (3.23) there exists in $\mathcal{B}$ a solution to system (3.20).

Proof

Let us consider in C the equation

$$\int_a^b \phi(x,y)[u(x) - u(y)]dy + [\psi(x,a) + \psi(x,b)]u(x) = g(x) \qquad (5.20)$$

Denote its solution for $g(x)=f(x)$ by $v_1(x)$, the solution for $g(x)=\psi(x,b) - \psi(x,a)$ by $v_2(x)$. The solution for $g(x)=\psi(x,a) + \psi(x,b)$ is the function $v_3 \equiv 1$. All these three solutions are unique due to the assumptions of this Theorem. Direct substitution shows that the function

$$v(x) = v_1(x) + \frac{1}{2}[u(b) - u(a)]v_2(x) + \frac{1}{2}[u(b) + u(a)] \qquad (5.21)$$

fulfils equation (3.20a) in the interval (a,b) for arbitrary values u(a) and u(b). We extend the solution (5.22) to an interval [a,b] choosing the values u(a) and u(b) in such a way, that the remaining equations of the system (3.20) be satisfied. The assumption on the absolute stability of the medium brings about that the difference u(b) - - u(a) is expressed uniquely by the functions v_1, v_2, ψ, ζ, p, q, while the sum u(a) + u(b) remains arbitrary. In this way a unique (up to a constant) solution continuous on (a,b) and bounded on [a,b] is constructed. Q.E.D.

Considering the above results we see, that the correct interpretation of the models of surface-volume type is as follows: they do not describe the body as a part of a medium, but as a system composed of two media: a volume medium and a surface one - the "skin" must not adhere to the body; it can be displaced with respect to the bulk interior. The case when $\psi(x)\equiv 0$ is here distinguished as essentially singular. Since no interactions between the surface and the interior media occur, the "skin" can arbitrarily be moved with respect to the interior of the body (the theorem on existence of a solution in $\mathcal{B}$ does not hold for this case).

In spite of the fact that the point (A) of Definition 3 is satisfied, we have

Theorem 4.8

The surface-volume integral model is not VS-compatible in $\mathcal{B}$.

Proof

The limit of the responses to a VS-sequence of loads is

determined by the interior bulk medium. In consequence, the limit solution u is not bounded at the boundary. On the other hand, according to the theorem of existence, the solution $\tilde{u}$ corresponding to surface loads is bounded. Hence, the conditions (B) and (C) of Definition 3 are not satisfied. Q.E.D.

4. Integro-differential models

We shall show that the system (3.24) is equivalent to the equation

$$\int_a^b \Theta(x,y)u'(y)dy + C(x)u'(x) = -p(x), \qquad (5.22)$$

in which the function p(x) defined on [a,b] occurs on the right-hand side.

$$p(x) = p + \int_a^x f(y)dy. \qquad (5.23)$$

Indeed, since from the condition of equilibrium (3.23) and from (5.23) it follows that

$$p(a) = p, \qquad (5.24a)$$

$$p(b) = q, \qquad (5.24b)$$

$$p'(x) = f(x), \qquad (5.24c)$$

then by putting x=a into equation (5.22) one obtains equation (3.24b) and by putting x=b one obtains equation (3.24c). Also by differentiating equation (5.22) one arrives at equation (3.24a). Thus, each (appropriate class) solution to equation (5.22) satisfies the system (3.24) and conversely.

Let us assume, that

$$C(x) \not\equiv 0 \qquad (5.25)$$

Then one can divide equation (5.22) by $\sqrt{C(x)}$ in order to transform it into Fredholm's equation. We have here

Theorem 4.9

If condition (5.25) is satisfied, then for each function f(x) representable as a derivative of a function continuous on [a,b] there exists in C^1 a solution u(x) of the system (3.24), unique up to a constant.

In what follows, we shall choose the value of the constant so that the solution vanishes at the point x=a.

Theorem 4.10

The integro-differential model is VS-compatible in C if condition (5.25) is satisfied.

Proof

Let us consider a VS-sequence $\{f_n\}$ on [a,b], for which

$$\int_a^b f_n(x)dx = F. \qquad (5.26)$$

We shall show, that the sequence $\{u_n\}$ of the solutions to equations (3.24) corresponding to the system of loads $\{-f_n, F, 0\}$ has the limit identically equal to zero.

Let us consider the sequence of equations of the form (5.22), where the right-hand sides are equal to

$$p_n(x) = F - \int_a^x f_n(y)dy, \qquad (5.27)$$

The sequence $p_n(x)$ is convergent to the function

$$\lim_{n\to\infty} p_n(x) = \begin{cases} F & \text{for } x=a \\ 0 & \text{for } x\in(a,b] \end{cases} \qquad (5.28)$$

which is bounded.

The solution to equation (5.22) can be represented in the form

$$u'(x) = -\frac{p_n(x)}{C(x)} + \int_a^b \frac{R(x,y)}{[C(x)C(y)]^{1/2}} p_n(y)dy, \qquad (5.29)$$

where $R(x,y)$ is the Fredholm resolvent of equation (5.22) divided by $\sqrt{C(x)}$. Due to the continuity of $R(x,y)$ and $C(x)$, the integral component in (5.29) tends to zero when $n\to\infty$. In virtue of this

$$\lim_{n\to\infty} u_n(x) = \lim_{n\to\infty} \int_a^x u_n'(y)dy \equiv 0. \qquad (5.30)$$

Q.E.D.

In the proofs of the theorems of existence we have made use of the condition $C(x)\not\equiv 0$ in a very essential way. In the opposite case, when $C(x)\equiv 0$, the results can be completely different. Let us consider an example with $a=0$, $b=\infty$ and

$$\Theta(x,y) = \frac{\alpha}{2}e^{-\alpha|x-y|}, \tag{5.31a}$$

$$C(x) \equiv 0, \tag{5.31b}$$

$$f(x) = F\beta e^{-\beta x}. \tag{5.31c}$$

Then

$$p(x) = p + F - Fe^{-\beta x}, \tag{5.32}$$

and we obtain the equation for $v(x)=u'(x)$

$$\frac{\alpha}{2}\int_0^\infty e^{-\alpha|x-y|}v(y)dy = F\beta e^{-\beta x} - (p+F). \tag{5.33}$$

The solution v we are looking for, can be represented in the form

$$v(x) = ae^{-\beta x} + b + cw(x), \tag{5.34}$$

where a, b, c are constants depending on the parameters α, β, p, F and w is a function to determine. This function should satisfy the equation

$$\frac{\alpha}{2}\int_0^\infty e^{-\alpha|x-y|}w(y)dy = e^{-\alpha x} \tag{5.35}$$

which, however, has no solutions in the concerned class of continuous functions increasing not faster than a polynomial. To show this, let us multiply equation (5.35) by $e^{\alpha x}$, and differentiate it with respect to x. As a result one obtains the condition

$$e^{\alpha x}\int_x^{\infty} e^{\alpha y} w(y)\,dy \equiv 0 \qquad (5.36)$$

which can be satisfied only by $w(x)\equiv 0$, i.e. the function for which relation (5.35) is not true. Thus, equation (5.33) has no solutions in the class examined, though it has solutions in distribution space.

Hence, the occurrence of a local term in the equations of statics is a significant condition, sufficient for VS--compatibility of the model.

5. Comparison between integral and integro-differential models

Comparison of the results obtained for the models considered shows some fundamental qualitative differences between the properties of integral models and integro-differential ones. However, a model of integro-differential type has been derived from the surface-volume integral model (Kröner[10]). The question arises as to what relation between the two categories allows us to pass from one to the other. Taking into account the contradictory properties of both models, it seems to be impossible. If we examine this relation more thoroughly, we arrive at the following equations

$$\phi(x,y) = -\frac{\partial^2\Theta(x,y)}{\partial x\partial y}, \tag{5.37a}$$

$$C(x) \equiv 0, \tag{5.37b}$$

$$\psi(x,a) = -\frac{\partial\Theta(x,a)}{\partial x}, \quad \psi(x,b) = \frac{\partial\Theta(x,b)}{\partial x}, \tag{5.38}$$

$$\int_a^b \phi(x,y)dy + \psi(x,b) + \psi(x,a) \equiv 0, \tag{5.39}$$

$$\zeta = \Theta(a,b). \tag{5.40}$$

These equations reveal an important fact: it is possible to pass from a particular kind of integral medium (equation (5.39)) to another particular kind (formula (5.37b)) of integro-differential medium.

For example absolutely stable models do not satisfy the relation (5.39). On the other hand, the condition (5.37b) implies VS-incompatibility of medium.

References

1. Rogula, D., On nonlocal continuum theories of elasticity, *Archives of Mechanics,* 25, 233, 1973.

2. Beals, R., Non-local elliptic boundary value problems, *Bulletin of the American Mathematical Society*, vol. 70, 693, 1964.

3. Weissmann, A.M. and Kunin I.A., Boundary-value problems in nonlocal elasticity, *PMM*, 33, 5, 1965 (in Russian).

4. Rymarz, Cz., Boundary problems of the nonlocal theory, *Proc. Vibr. Probl.*, 4, 1974.

5. Sztyren, M., On boundary forces in a solvable integral model of nonlocal elastic half-space, *Bull. Acad. Pol. Sci., Ser. Sci Techn.*, vol. 26, 299, 1978.

6. Sztyren, M., Boundary-value problems and surface forces for integral models of nonlocal elastic bodies, *Bull. Acad. Pol. Sci., Ser. Sci Techn.*, vol. 27, 327, 1979.

7. Sztyren, M., Boundary-value problems and surface forces for integro-differential models of nonlocal elastic bodies, *Bull. Acad. Pol. Sci., Ser. Sci Techn.*, vol. 27, 335, 1979.

8. Kunin. I.A., Theory of elastic media with microstructure, *Nonlocal theory of elasticity*, (in Russian), Moscow, 1975.

9. Datta, B.K., and Kröner, E., Nichtlocale Elastostatic: Ableitung aus der Gittertheorie, *Zeit. Phys.*, 196, 203, 1966.

10. Kröner, E., Elasticity theory of materials with long--range cohesive forces, *Int. J. Solid. Struct.*, 33, 1967.

11. Edelen, D.G.B., Nonlocal variational mechanics, *Int. J. Engng. Sci.*, 7, 269, 1969.

12. Eringen, C., Nonlocal polar elastic continua, *Int. J. Engng. Sci.*, 10, 1, 1972.

13. Rogula, D., and Sztyren, M., On the one-dimensional models in nonlocal elasticity, *Bull. Acad. Pol. Sci., Ser. Sci. Techn.*, 9, 341, 1978.

14. Rogula, D., and Sztyren, M., Fundamental one-dimensional solutions in nonlocal elasticity, *Bull. Acad. Pol. Sci., Ser. Sci. Tech.*, 10, 417,1978.

15. Barnett D.M., Om admissible solutions in nonlocal elasticity, *Lattice dynamics*, Pergamon Press, 1965.

Printed in the United States
By Bookmasters